Military Logistics

The Third Dimension

Military Logistics

The Third Dimension

Brig. Parmodh Sarin (Retd.)

Manas Publications
New Delhi -110002

MANAS PUBLICATIONS
(We Convert Fighters into Writers)
(Publishers, Distributors, Importers & Exporters)
4402/5-A, Ansari Road (Opp. HDFC Bank)
Darya Ganj, New Delhi-110 002 (INDIA)
Off. ✆ **23260783, 23265523, Res. 23842660**
Fax: 011-23272766
E-mail: manaspublications@gmail.com
manaspublications@yahoo.com

2014 (FP - 2000)

ISBN 81-7049-090-1
₹ 795/-

Designed at
Manas Publications

Printed in India at
Rajdhani Printers, Delhi and published by Mrs Suman Lata for
Manas Publications, 4402/5-A, Ansari Road,
(Opp. HDFC Bank),
Darya Ganj, New Delhi-110 002 (INDIA)

Contents

Contents

Preface

When I was a cadet and when things like 'verbal orders' were being taught, after a detailed briefing on the operational side of the problem, the administration aspect was covered in less than one sentence - 'bandobast normal'. This, inter alia, implied that somehow all the administrative requirements will be met and there was no real need for us cadets (busy as we were) to expend time, energy and effort in learning the intricacies of material support necessary for any kind of operation. That we were not to worry about anything concerning the logistics, and that, somehow we would get what we needed to achieve success. How wrong this all was I learnt the hard way. Fact of the matter is that an operational commander cannot divorce himself from the administrative affairs because in the complexity of modern war, he cannot operate independently from his support. That the logistics support hasn't as yet got its rightful place in our thinking has been goading me for years to present the case on behalf of the "support services". I started giving a thought to writing on the subject of logistics when I did a dissertation on the theme of a "National Logistics Grid"— where both the Military and Civilian needs are enmeshed into one—for my Master's degree in "Materials Management". My experiences over the years when I had felt concerned with my combat arm friends taking their logistics lightly also provided a strong impetus. When I started doing necessary research on the subject so as to strengthen my rationale with regard to various issues and aspects like "defence and (some call it versus) development, the economics involved in equipping the military, I found that very rarely has a complete view of the whole thing been taken.

Originally, I wanted to put across my point of view on the subject of logistics support to combat forces on the lines of Major General Julian Thompson's work *The Lifeblood of War- Logistics In Armed Conflict*. However, I felt that in taking that approach I was severely handicapped. The General Officer could draw on specific lessons from numerous operations where the American forces had actively participated- like Korea, Vietnam, and more recently in Kuwait where the massive logistics exercise was essentially a high-tech affair. We in India really haven't had that kind of experience and exposure for one to discuss the issues involved in a familiar environment. More importantly I wanted to project the case for logistics to be put on the same platform as 'strategy' and 'tactics'. Not surprisingly as I went along the canvas kept widening. To convince commanders that logistics is the third dimension of their strategic and tactical plans has been my main charter. That national industry, economic situation, technological base, and the will of the people are important to a well-equipped military has been my focus. Yet, despite this ambitious scope, I am uncomfortably aware of certain limitations, of further issues that might have been raised, or issues that might have been treated more comprehensively.

In my endeavour to study the user requirement and in order to put across a workable solution I have drawn on the experience of others. Therefore, it would be wrong for me to claim that all the ideas and concepts discussed herein are mine. They have basically stemmed from my discussions with numerous professional colleagues. From what I have been taught and studied in the field of logistics, from what I have practised over the years and from what I have seen being practised by my more competent peers and senior officers of the Support Services. I hasten to add that all the opinions and theories vis-a-vis a more proficient and efficient logistics service as projected in this work are generally being practised in certain more advanced armies of the world. Or these are being examined as lessons learnt from the 'Kuwait liberation' or as after effects of the break up of the Soviet Union. Since my experience has largely been confined to the 'Army Ordnance Corps', I have generally taken the 'material support' aspect to the army as my main theme. I think I know that in so far as Navy and Airforce are concerned their problems are the same where these relate to logistics.

It is always difficult to assess what a book's readership will be. In writing this book, I have had hopes that it will be of interest to both the specialists and students of the subject of logistics. Be it the military, para military or commercial establishments wanting to or capable of making a meaningful contribution in the arena of material support to the Services they will hopefully get an adequate view of the issues involved. I also hope that my work will be of interest to the people who are averagely informed about 'military logistics' and many dimensions of the problem. And to those who always complain (wrongly - you will agree as we proceed with the narrative) that too much money is being spent on the military. There are bound to be divergent views on some of the suggestions that I have advocated regarding solutions to our problems in providing an efficient material support to troops. Where a thinking Military is concerned that is how it should be. I claim no "final solution" to our problems.

Logistics principles (as enunciated) are often masked by a seemingly impenetrable veil of jargon. I have tried to strip away that veil by explaining the complex field of logistics in a reader-friendly language. My aim in putting across the intricacies of the logistics support system is to make combat commanders more pragmatic and aware of the limitations and capabilities of the support services. My endeavour has been to share my experiences and views on the intricate subject of "military logistics" with the users and the providers of this support. With these aspirations towards an expert and a general readership, I hope the balance of the language and presentation of arguments will satisfy both. I have tried to avoid boring the expert with long winding explanations of familiar terms, scenarios and concepts. And to avoid baffling the lay reader with esoteric and confusing tables, charts and graphs, I have tried to retain a somewhat smooth expression by generally excluding them. Some experts in the field may feel that with 'betas' and 'sigmas' I could have made the arguments more convincing. But my "ultimate" target readership is the operational commander and not exclusively technical specialists in the wide range of fields associated with logistics. The level of commanders can range all the way from a company commander up through the chain. And if I cannot grab their attention then this is a wasted effort.

Brig. Parmodh Sarin (Retd.)

Acknowledgements

A large portion of practical suggestions offered herein have been gleaned from an array, though limited, of research and scholarly works on the 'unglamorous' subject of logistics support to a modern military force. Numerous proposals and ideas concerning various facets of the task have generally emerged from this reading. I am grateful to Major General Nirmal Sondhi (Retd.) AVSM and Brig. Y Mukherjee SM for having gone through the manuscript and for their practical suggestions in order to enhance the readability and utility of this book. My special thanks are due to a large number of friends who provided encouragement and help while clearing my own mind about certain activities concerning quality of the logistics service provided to the combat troops. If I list them individually I might miss out on some; therefore I prefer to be generic.

Unless, specifically identified as the source, all the opinions expressed herein are mine.

Introduction

Military history abounds with the dramatic fare of battles and campaigns, but definitive analysis of administrative functions is a relatively neglected field. There are a number of studies, both serious and anecdotal, on every military campaign right from the ancient times. Unfortunately, there are very few on the subject of resources required in providing material support for those very operations. Maybe, it is so because commanders better known for their strategic and tactical brilliance while recording their exploits have generally just glossed over the effort involved in making them achieve the successes that made them famous. Conceivably, because logistics do not have the seductive and engrossing power of operational thrusts, deep penetrations and swings in various directions. It is basically all about tonnage and load carriers, maintenance and repairs, casuality evacuation, and rations. Such details do tend to slow down the readability of the literary recordings of history-making events. To be fair, some of these commanders have, at times and on hindsight, lamented not having paid enough attention to the logistical aspects of their operational plans. In terms of resources, time required to make the facilities available and the material inadequacies they preferred to either ignore while launching their campaigns or had thought that somehow everything will work out. They do make a mention *(en passant*, if you please) of the men and opportunities they lost for having defaulted in not giving appropriate importance to their logistics support or for having taken it for granted. To the dead and the wounded in the campaigns this belated realisation did no good because it normally came when the operations having got stalled for want of either fuel or ammunition, caused heavy loss of life and limb besides causing untold misery to the soldiers and civilians alike.

One could fill a whole volume with quotable quotes from the memoirs of various commanders right through history, where they have fretted (generally after having got grounded by the enemy) over the lack of logistical appraisal of their operational plans. They have even lamented about the material inadequacy of the country's system to support those plans. There are some instances where, on hindsight, the lack of ammunition or fuel for the tanks, or, an inadequate 'medical cover' and insufficient rations were eventually identified as reasons for retreats or halts in the operations. Often enough this halt in the campaign allowed the enemy to regroup and wrest the initiative. If any discernible lesson vis-a-vis logistics got learnt in the process, it is usually not very clear. Perhaps, overall success of these illustrious commanders led to a thinking that in the ultimate reckoning it did not really matter whether the logistics were given enough thought or not. Rightly, there has been, in recent years, an increasing interest in military logistics.

Some economists- not necessarily from India- also argue (reasonably so) that defence is a better stimulus for the industry, because it is more capital intensive and creates a greater economic multiplier for all the rupees invested therein. It can bring in an element of urgency where development of a new product is concerned and can galvanise a manufacturing base for a wide range of goods and services. It will be my case to establish that the logistics needs of the military bring in development. And that they do give a boost to the national economic well being provided a holistic view is taken at an appropriate level. Economic strength and industrial capability serve as the basis from which all logistics activity must start if workable and sustainable operational plans are to be made. For, it is that foundation on which the government and the military leaders must determine the ability of existing and foreseeable logistics assets to support military strategy because a successful strategy is not simply a good plan; it requires adequately equipped forces to carry it out. The degree of a nation's economic strength determines the amount of resources it can make available to logisticians in responding to a security threat. For all the wrong reasons we have always considered expenditure on defence as a 'non plan expenditure' thus denying ourselves a chance to develop an industrial and technological base which could provide us with a higher level of industrial and economic growth along with superior defence. One misconception that I have tried to dispel is that 'defence and development' are not complementary, and that they are disputants in so far as utilisation of the country's resources is concerned. A very limited discussion on this aspect of growth, our inhibitions in involving the private sector in meeting military needs, and economical logistics support has formed part of this analysis of "logistics".

As it happens in all democracies there are any number of experts, homegrown or otherwise, who always bemoan the fact that too much money is being spent in meeting the defence needs. They seem to overlook the fact that the usual state of man- and of nations - involves competition, and in an expanding world it actually means a struggle for existence, dominance and breaking into the growing markets and hence a conflict of interests. We see a continuing conflict between peoples and between nations, and we see that there is no real peace in our generation because the conflict is unceasing. It is a conflict that at all times involves violence; and this violence may be political, economic, diplomatic, military, para-military or "State sponsored terrorism". This conflict is simultaneously formal and informal, it is ideological and physical, and it is within nations as well as among nations. We live in an age of great insecurity. At the lower levels the sources of insecurity range from terrorism and insurgency through national radicalism and religious fanaticism; at the upper levels, they include hostile neighbours. We, as a nation, have been going through all this in the form of low-intensity conflicts and proxy wars for a long time now. It will be my endeavour to dispel misgivings of those experts who in their mistaken belief in the goodness of human nature think that a country of our size could wish away the problems of national defence, or who think that there could be cheaper alternatives. To maintain and sustain the integrity of the country we have to expect a huge expenditure to be incurred both in the form of manpower and resources; there is no escape. Defence expenditure has to be put into the context of the environment in which a country has to exist in the real world. The external and internal aspects of the environment in which the military is expected to perform its designated tasks serve as a framework for this analysis

Military logistics cost a lot of money, something that a developing country like India could put to a more constructive use. In our environment the 'bottom line' is that an important aspect of an efficient material support for the military has to be cost effectiveness, and it must relate not only to economic costs but also to the growth and progress of the nation. Blending our civilian capability with our military needs can help us achieve a higher level of material support without undue strain on the national economy. It is proposed to focus some attention on this aspect. It is also proposed to make a case that the logistics requirements of the Services actually lead to industrial and technological growth because the industrial might and the technological strength of a nation are intimately connected with its military might. I have tried to deal with the subject both as one committed to providing material support to the front-line soldier and as a tax-payer who would like to see that the heavy taxes that he pays and the country's resources are channelised in a direction where the needs of the military and aspirations of the common man actualise. Because of the environment of 'resource crunch' and therefore, somewhat half hearted efforts at improving our logistics support to the frontline soldier an ever increasing gap between the military's need for materials and their availability has crept in.

The reality of and potential for the domestic use of armed forces against political and other groups who oppose the state in one form or the other tells us that our traditional support in the vertical chain is no longer effective. This also means that compartmentalised material support to army, navy and air force is no longer economical or technically viable unless some efficiency is compromised. More importantly, there is a great deal of duplication of effort in the logistics organisations of the three services. Consequently, there is a great wastage and inefficiency in the operation and functioning of the services. The time has come for us to think in terms of 'integrated logistics' and may be for the creation of a "Defence Logistics Service". In spite of my intention to do so I have been able to give only peripheral attention to urgent requirement of a national level logistics policy- a subject with tremendous potential for a detailed study in its own right. For that matter, because I wanted to cover all my flanks where as material support to the troops is concerned no special treatment has been possible to a few serious and debatable issues. I have, however inadequately, tried to make a case for a "Logistics Corps" as a first step in this direction. Common denominator in this treatise on the subject of logistics is that all commanders want a reliable and efficient material support both in peace and war. My mission has largely been to tell commanders that unless logistics support their plans they have no reason to hope for success. My thesis is that commanders have to accept the fact that any military plan, if it is to be realistic, it must start with logistics considerations. And the logistics plans, in turn, must be interwoven with national industrial, economic and development plans. That the changing dimensions and perspectives of warfare at the strategic levels must be recognised in logistics support. It will also be essential to realise that strategy and tactics are always limited and are, necessarily determined and influenced by the 'third dimension' ie, logistics factor. Accordingly, it is proposed to highlight that Strategy, Tactics, and Logistics are three sides of an equilateral triangle. Any imbalance can have serious repercussions on the country's defence preparedness.

One big problem facing the commanders today is that the requirement of technical support, ammunition and inventory has multiplied exponentially, with little or no "scientifically evaluated" consideration being given to the force level of the 'tail element'. Fact of the matter is that our logistics organisation, which was designed during the Second World War, is not in tune with the present times. It seems to be over-structured and too imbalanced towards functionalism, with too little weapon systems or commodity orientation. Logistics organisations must change and grow and develop compatibility with the fighting organisations as per the requirements of a particular theater, the range and peculiarities of the equipment deployed and the engineering support needed to keep the weapons and other resources in battle-ready condition. 'Low intensity conflicts' and military's prolonged deployment 'in aid to civil authority' have become regular features of a combat soldier's professional life. Our traditional methods of material support having become outdated it is time we gave thought to "horizontal logistics support" where a soldier could draw his essentials from any service installation, he does not have to be dependent on a particular establishment for this purpose. The subject of "inventory automation and asset vision" has been discussed at some length.

Where a combat commander is concerned, logistics itself has little purpose other than to create and to support combat forces and logistics should be responsive to the needs of the command. No matter what we, in the services, do there is always an uncovered 'logistics' flank. And on top of it commanders always produce arguments to reduce the "tail" while assuming that the level of support will remain the same. So that their operational plans prepared so diligently and carefully and practised exhaustively need not suffer from inadequacy of resources, be it in material or technical manpower an effort to put across the case for a better understanding of the "support services", has been made. It is proposed to convince these commanders that having effectively moved into an era of technology-driven machine warfare, only in extreme moments will the individual warrior ever again play a pivotal role. I have, as an officer from a supporting service, elected to treat ' tail and the teeth ratio' as a separate chapter in the fond hope that the operational commanders may eventually agree that there can be no teeth without an appropriately sized tail.

In this book I have tried to draw attention to a number of issues related to logistics and what we should be doing to enhance our capability in so far as logistics support to our forces is concerned. I have put forward a few suggestions for improvement of our material support within the framework of the national capacity and capability. It would be surprising if all of them found ready acceptance. No set of theories, however persuasive, can by themselves obviate or reconcile differences of opinion. I have tried to make a case for a strong logistics base at the national level drawing heavily on the experiences of democratic forces. My intention is to promote constructive discussion on the subject with a view to generate a consensus and meaningful measures where the heavy investment the country makes in equipping its military is concerned. The combat soldier must have faith and confidence in the system that is required to ensure his battle-worthiness. Since that would be the aim of both the combat and combat support commanders, there is neither any room, nor any intention for creating any controversy.

Glossary

ABC Analysis ABC analysis - This is related to the annual usage cost of a particular item. 'A' items cost approximately 60 to 70 % of the budget, and they are lesser in number. 'B' items cost 20 to 30 % of funds, 'C' items are always in a larger number but cost hardly 10% of the money available.

Accessory In addition to a complete article such as the spare wheel of a vehicle without which the item can be used.

Area of Supply. An area allotted to a service unit, all units in that area are dependent for the supply of stores.

Assembly Composite components commonly supplied ready assembled, such as a recoil system of a gun.

CADS Containerised Ammunition Distribution System

CONCOR Container Corporation of India

Census Stores Items that are vitally important and whose supply is restricted necessitating a periodic return in order to examine the overall stock situation Vis a Vis troops requirements.

Class 'A' Stores Usually main equipment, the liability for which consists of the troop's entitlement, maintenance requirements and policy reserves as defined by the 'general staff'.

Class 'B' Stores Items related to main equipments and items of general stores and clothing. The provision is based on scales or issue experience.

Controlled Stores These are those items where the issue is controlled by the Service Head Quarters. A control is imposed due to importance, high cost, or difficulty in procurement. An item can be decontrolled depending on improved supply position or any other reason. An important facet is that all the controlled stores are also "census stores", but all the "census stores" need not be controlled.

CONEX Container Exchange

DGQA Director General of Quality Assurance

DROPS De-mountable Rack off-loading and Pick ups

Dues In Stores due for delivery to a logistics installation in response to issue instructions, demands on base or other depots, DGOF, or from abroad.

Dues Out. Stores owing to a unit or establishment from a Service Installation and are not readily available for issue.

Expendable Stores Items, that lose their identity or are consumed in the process of repair, maintenance, packing, or manufacture of an article.

FSN Analysis- 'Fast, slow and normal', determine the consumption pattern of an item

IWT Inland Water Transport System

ISG Initial Scaling Guide

ISPL Illustrated Spare Parts List

Issue-in-Lieu When an item can be issued instead of another but provision for both is required to be made separately, the item is classified as issue -in-lieu.

MLRS Manufacturer's List of Recommended Spares

LDCs Less Developed Countries

LSA Logistics Support Area

Mobilisation. The term is applied to the process by which an armed force passes from peace to a war footing. The mobilisation of a unit entails its having full complement of its personnel, vehicles, stores, materials, weapons, ammunition etc.

Obsolescent Stores An equipment/ store for which no further provision will be made but the existing stocks, if any, will be used till they are exhausted.

Obsolete Stores Equipment for which approval has been given for its withdrawal from Service.

Primary Items: Primary items could be understood as those items that are necessary for the organisation. In this, weapon systems like tanks, APCs, guns, missiles, aircrafts, guided weapon systems, ships, submarines, and other systems like communication system, radar systems, vehicles etc., could be considered as 'primary' items.

RSI Rationalisation, Standardisation and Interoperability

Repairable Item Repairable item is an item that can be repaired or overhauled at least once and generally a number of times. These items are distinct from consumable items which get consumed either in the process of normal maintenance or after usage once a consumable item is discarded. The repairable item passes through a cycle which in its simplest form has two parts- in service and undergoing maintenance. Maintenance will also mean repair and overhaul.

Repair Pool. The amount of stock (serviceable and repairable) that is temporarily non-effective for having gone to workshops for repairs. Such an identified pool is always maintained, both to enable advance replacement issues to be made and to ensure even flow of out-turn from workshops.

SAP Analysis -'Scarce, available, plenty' allows us to build into the provision forecast the 'Availability State' of an item. Quantity to be ordered will also be governed by the 'scarcity' factor.

Secondary Items: Are those items that are necessary to keep the primary items in a state of serviceability and operational readiness. In this category come spare parts, assemblies, raw materials, fuels, lubricants etc. Such items remain idle until they are required to be used /fitted in an equipment, as distinct from the equipment themselves which are in use. Such secondary items are called 'inventory' and are managed by using various techniques of inventory control.

VED Analysis- This relates to vital, essential and desirable status of spare parts, clothing items, rations, medicines and so on. As the terms imply, certain parts and assemblies are considered to be vital for meeting operational requirements.

War Wastage Reserves (WWR) Stocks held for specific war purposes. Reserves of different items are held at various locations depending on the perception of operational commanders.

"If manpower and military hardware constitute the vehicle, which applies the necessary force to win a war, logistics are the power plant and fuel, which drive it." Logistics is the physical manifestation of warfare.

Dedicated to the unsung heroes of the "Support Services"

1

In Retrospect

During the 'Senior Command' course at the "College of Combat", Mhow, I often felt alarmed when some well-meaning and competent Infantry and Armoured Corps officers felt it to be below their dignity to talk of logistics. The material support for the operational plans they so painstakingly made for the 'sand-model exercises' was of extremely limited relevance where they were concerned. As a sort of concession to their peers like me, at best, they were willing to concede that such things like "logistics" and the men providing service did exist. "I have my Ordnance, EME, ASC, Medical, and other support services representatives to look into various mundane problems concerning battery, tyre, tube, ammunition, repairs, rations, and so on. Where is the need for me, the 'operational commander,' to worry about these insignificant things?" was the usual refrain. The implication was that the thinkers studying strategy find the drudgery of nuts and bolts unstimulating, meaningless, and wasteful. It is just an unfounded and unsubstantiated notion that any serious study of logistics is, by definition, plodding, dreary, technical, and unimaginative. I did not see any malice in this view.

According to Clausewitz, even Napoleon was originally not particularly inclined to listen about logistics. Whereas at one point his orders were "let no one speak to me of provisions", he eventually came to the conclusion that an "army marches on its stomach". (Field Marshal Rommel, the leading figure in the first desert war between two mechanised armies, was asked by the German Chief of Staff how he planned to supply the two additional panzer corps that he was then demanding from Berlin, for an offensive few in Berlin wanted him to mount. " That's quite immaterial to me," Rommel replied coolly, " That's your pigeon"). How he learnt his lesson is a fact of recorded history. For, he was to say later " the battle is fought and decided by the quarter masters, before the shooting begins." These homilies were simply wasted on my 'Combat Arms' friends on that course.

I did try to bring to the notice of those future commanders that the probable terrain of operations, the capability of the enemy to muster his logistics in the area of operations does have a bearing on the ultimate outcome. I also tried to reason that for a "mountain warfare exercise", a commander must take into account the fact that his troops must have clothing that provides warmth in extremely cold climates, their food must be capable of being easily prepared at high altitudes, provide warmth and nourishment, and their weapons must be light and capable of functioning efficiently throughout the wide temperature range. Only then the commander can hope for success. Similarly, peculiarities of every terrain and specific material requirements that must be met, be examined alongside the tactical plan. My efforts to draw their attention to the aspects of material needs for those plans and the operational commander's role in ensuring that he had the wherewithal to launch his operations was lost in the urgency for detailed consideration of the tactical situation only. I would (during that course) often give up in despair.

An officer from the services is not expected to have any serious exposure to real life operational situations, they had their reasons for not taking me seriously. Such experiences are not uncommon in any army of the world when advice of the logistician is either not sought or ignored in a hurry to work out the 'tactical plan' that finds acceptance of a higher commander. I have always felt that rather than a lack of knowledge the problem may lie in a reluctance to acknowledge the importance of logistics in war. I have usually found that the commanders in their hurry to measure up to operational commitments tend to either ignore their material requirements or take these for granted. To my mind such an attitude is dangerous. It assumes and wrongly, that the crises are met and wars are fought by the 'elite' strategists. And the logisticians are uncomfortable, unwelcome meddlers in all but the most mundane aspects of policy strategy. Logistics is traditionally an unglamorous and under-appreciated activity. And when did the subject of logistics ever have enough 'sex appeal' for the bright and upcoming combat commanders of the future to get seriously involved in? Till date I find that a genuine understanding of logistics, at whatever level of command, is an exception rather than the rule. Maybe that is why many of our unresolved problems concerning equipping, sustaining and maintaining the troops remain logistical. There are still deficiencies and contradictions in our logistics systems. While some of these deficiencies are caused by our failure to properly apply what we already know, there are many areas where we are limited by our imperfect knowledge of the art and science of logistics. Our military schools still do not teach resource management as part of their curriculum. Consequently, military planners largely overlook resource management in the course of developing and conducting war strategy. This oversight may result in commanders developing overly ambitious strategies that cannot be supported by the limited resources available and, possibly, dooming the strategies to failure.

It is rather disconcerting to note that most of our peace time exercises make little or no pretense of having any realistic consideration of the logistics effort involved in making those plans work. The usual refrain always has been that to do so would distract the 'trainee commanders' from totally, and effectively grasping the intricacies of the tactical situation, figuratively described as their 'bread and butter', not realising that both bread and butter are important facets of logistics. Another reason given is that in addition to tactics and strategy, consideration of the logistics is likely to unduly complicate the manoeuver being practised. Consequently and wrongly, we continue to train our future commanders with hardly any importance being given to the 'life-blood' of every campaign. Our endeavour to make them more efficient operational commanders may, in all likelihood, become counter-productive if they are not made aware of the complexities of making material requirements available for their operational plan. In its relation to strategy and tactics logistics assume the character of a dynamic force, without which the strategic conception is simply a paper plan. Strategy and logistics must proceed together as preparatory concepts and activities to create viable tactical packages because both strategy and logistics are essential for planned tactics. Specific material requirements that must be met need to be examined along side the tactical plan. Thereby meaning that success or failure, be it a proper war, a proxy war, or a low intensity conflict, is directly related to making and practising those plans which are logistically viable. Often enough, one sees operational plans being made with inadequate analysis of the material support required to make those plans work. The economic viability, both in terms of men and material, of such plans also sometimes does not get the attention and concern it deserves.

An operational commander (particularly one who is honing his tactical skills) must realise that a changing operational situation often demands that the material support also alters its focus. Any such changes if they involve major efforts soak up scarce logistic command and control assets and demand extra managerial attention. We seldom realise that in the absence of adequate and appropriate resources, and where the logistics do not support operational plans, such plans are bound to fail in any case. Therefore, these peace-time exercises conducted so painstakingly and 'ostensibly' lacking in appraisal vis a vis material support consisting of huge tonnage of ammunitions required to be moved, preserved, cared for and delivered to the combat forces, thousands of liters of all kinds of oils and gasses, the rations and all important medical care, create a wrong impression in the minds of commanders. They get away with the delusion that their tactical brilliance will, some how, compensate for their inadequate consideration of logistics. Consequently, fairly complicated plans generally get made and rehearsed. And every body goes home happy, having done an extremely brilliant job of producing an appealing plan, which, due to the usage of 'multi-coloured' pens, a lot of arrows and symbols appeared to have found favour with the higher commander. This results in a situation where an important aspect which the commanders must learn to respect, that is, the fact that complicated operational plans breed numerous logistic areas and focal points prone to enemy actions and thus harmful to operation in hand, does not get registered. This is an oversight that could have serious consequences in case these plans ever get to see the light of the day. To achieve the operational goals commanders will have to associate themselves with the logisticians all through, so that based on professional and expert advice, changes where required, can be made well in time. Commanders must accept that logistics plans are vital, they are ambient and they are all pervasive, so as to be considered to be the common denominator of all plans.

Preparing forces for deployment requires an unprecedented degree of insight and forethought, and actually deploying them raises a wide array of logistics issues that often are left unanswered in these peacetime exercises. These issues are often given only superficial consideration even while considering troops and their day to day maintenance in real situations. We must remember that, all operational plans are normally paper plans and are therefore, often enough bold, innovative, and are suitably guided by the wisdom of the genius of the past. These plans reflect the strategic brilliance of the commander. He can afford to be creative, unconventional, forceful and "Napoleanic" in his approach. Because these plans do not involve any specific expenditure until the emergency begins, a commander can afford to be expansive and original. Logistics plans on the other hand demand immediate translation into resources in the form of skilled manpower, real estate, materials and money. Naturally, they take time to fructify depending on the kind of a specific technology required and the time and effort involved in making the logistics support available for actualisation of those operational plans. The end user also has to be made familiar with a particular product if it is a new introduction. It is important to understand that the logistics needs of the military often have a long gestation period. In today's environment a number of predictable and unpredictable uncertainties will make it even more important for the operational commander to thoroughly understand the logistics aspects. Then only will he be able to make a rational assessment of his capability to undertake any operation.

There have been some improvements in our strategical, tactical, and technological concepts and doctrines. But, where as logistical needs of the military are concerned, we have, by and large remained confined to the thinking of the olden days. Perhaps, the limitations of resources both in the form of funds and technology have restrained the decision makers from being pragmatic, leading to a situation where we have made very few and 'largely' cosmetic changes in our perceptions where equipping our military is concerned. It is not that other armies in more advanced countries do not have these problems. They all have their limitations in the form of financial, technological, and trained manpower. In our case a major and an additional restraining factor is that we are too rule-bound and traditional in our approach to the all important task of creating a dynamic and forward thinking war machine. To establish our credibility as a nation that can look after its integrity we have to identify our priorities where near-term readiness is the top priority, mid-term modernisation is the second, and long-term sustainability is the third priority. In the context of affordability, we must adequately prepare to fight the initial phases of war regardless of its ultimate duration. The operational commander has to take into account all these aspects when he trains himself for "higher direction of war".

In order to have a well oiled war machine we need to have an up-to-date economically and industrially viable and competent administrative back up to the tactical organisations. This means a well-developed Research and Development facility, smooth and reliable communication system, and an intelligence system that gives relevant inputs to the planners well in time. There is also an apparent need for periodic evaluation of various factors like geopolitical situation, technology, and state of relationship with neighbours, as these along with the economy have a direct bearing on the national foreign and defence policy. The type and level of industrial development available in the country will also have a direct bearing on the assurance levels vis-a-vis logistics support to combat formations. Our operational plans will invariably be influenced with raw materials required for a particular system and their availability in the country. This careful evaluation is particularly important in connection with those organisations and procedures that were established by the British under wartime compulsions and to suit their convenience. This will help reap full benefits from the huge investment the country makes in the form of men, materials and money. While equipping its forces policy makers must periodically look at the efficacy or otherwise of the Services. They should also concern themselves with what all the military can do to make logistics support more efficient on ground without increasing the expenditure. This will help in taking a 'course correction' before it gets to be too late and therefore a 'fait accompli'.

2

What is Logistics?

The military activity known as logistics is probably as old as war itself. The word 'logistics' is derived from the Greek adjective "logistikus" meaning "skilled in calculating". Research indicates that the first use of the word with reference to an organised military administrative science was by the French writer Jomini who served as a staff officer in Napoleon's army. In 1838 he set down logistics as one of the six branches of the military art, the other five being statesmanship in its relationship to war, strategy or art of properly directing masses upon the theater of war, grand tactics, engineering, and minor tactics. He included the phrase 'it is the execution of strategical and tactical enterprises' in his definition of logistics. In short, he devised a theory of war upon the trinity of strategy, ground tactics, and logistics. He defined it as "practical art of moving armies". Various campaigns over the years added numerous facets to the term 'logistics'. In early 20th century, logistics began to be viewed as one of three sub-divisions of war-making, namely, strategy, tactics and logistics. Net result was a comprehensive definition in the year 1948 in the " Dictionary of United States Military Terms for Joint Usage" that describes logistics as "... that part of the entire military activity which deals with planning, production, procurement, storage, transportation, distribution, maintenance and evaluation of personnel, supplies and equipment; with induction, classification, assignment, welfare and separation of personnel and with facilities required for the support of the military establishment, including construction and operation thereof. It comprises both planning and implementation". Over a period of time it was found that the canvas was too large and virtually unmanageable. Therefore, based on experience gained over the years the term was redefined in 1968 to say " logistics is the art and science of creating and maintaining a military capability. It consists of the process of determining requirements, acquisitions, distribution and maintenance of materials".

Our manuals have a somewhat restricted definition, in that we have confined ourselves to accept that "logistics is the science of planning and carrying out the movement and maintenance of forces." In today's usage logistics is the function of providing all the material and services that a military force needs in peace or war". One can say that broadly all activities other than strategy and tactics can be covered under the heading of logistics. Since provision of services and material cost a lot of money and resources, a nation has to find adequate funds, firstly to equip its forces, second to maintain and sustain them for an indeterminate length of time. The country also has to undertake regular upgradation of military technology, improve the quality of services, and yet remain within its means if the military is not to become a 'white elephant' In short, logistics can be described as the bridge between our combat troops and the industry and natural resources of our country. Our objective should be to widen the bridge and shorten its span if we are to remain within our means. Logistics is not an exact science. It is never easy to lay down a prescribed limit of resources that can be devoted to sustain our forces. There is no mathematical formula or set of "tables" which tells us precisely what supplies or services will be needed, where and when they will be needed, or the best way to provide them. In its widest sense, logistics comprehends all functions relating to the supply of the means to conduct military operation and could thus extend to the mobilisation of the national industrial, technological, economic and human resource base. One can say that the national production capacity is directly linked with its military strength both in the times of peace and war. The U S 'Joint Chiefs of Staff' committee have defined logistics as: Design and development, acquisition, storage, movement, maintenance and distribution of material. Movement, evacuation and hospitalisation of personnel. Acquisition or construction, maintenance, operation and disposition of facilities. Acquisition or furnishing of services.

Doctrine describes logistics as a process that consists of planning and executing the movement and sustenance of operating forces in executing a military strategy and operations. It has a direct bearing on a country's capability to support a national strategy. To put the matter in its right perspective one can say that "Logistics is the planning of and providing goods and services to the entire population of the country. The levels of provision of the goods and services will be governed by the country's 'Gross Domestic Product', and the capability of its industrial and technological base to ensure an appropriate level of provisions". There will be different compartments dealing with different sections of society. These are:

- National logistics is the process of planning for and providing goods and services for the support of a nation's military forces and its operation. A nation's civilian economy, its international obligations and the material requirements of its population will be major factors in deciding the level of expenditure that can be incurred in sustaining the determined force level.
- Civilian logistics is the process of planning for and providing goods and services for the civilian population, its economy and perspective for growth. It also means mobilisation of the civilian industrial economy to sustain the armed forces both in peace and war. Naturally, this may cause some strain on the common man who may feel that his right to a better life is

getting somewhat restricted due to military usage of goods and services. For, armed forces are not the only ones who need goods and services. Other than weapons, ammunitions and certain specific items that have only military usage, the whole country has these needs. The important thing is to identify the areas of commonality. This exercise is dominated by civilian influences and by civilian authority. An authority that may or may not appreciate that a major criteria for military logistics proficiency is all round production efficiency.

- Military logistics is the process of planning for and providing goods and services in the support of military forces. This would take into account a country's overall infrastructure including the trade and its industry. The planning would include the long term and short term perspectives, strategic goals, and overall national capacity to absorb the expenditure on account of sustaining a military force.

where as the military is concerned, logistics is the everything else (the tactical brilliance etc., being apart)- art and science of warfare. Logistics include hundreds of possible functions since everything needed to support warfare could form part of this list. Logistics is essentially moving, supplying, and maintaining military forces. It is basic to the ability of armies, fleets, and air forces to operate—indeed, to exist. It involves men and material, transportation, quarters and depots, communications, evacuation and hospitalisation, personnel replacement, service, and administration. In its broader sense, it has been called the economics of warfare, including industrial mobilisation, research and development, technology, funding, procurement, recruitment testing, and, in effect, practically every thing related to military activities besides strategy and tactics. Today more important is the fact that the operational commanders familiarise themselves with all the aspects of logistics. Where the logistician is concerned, the basic logistics framework should be such that it is able to support alternate, different, and often, bold and innovative strategic and tactical plans. That all such plans should be within the national capabilities be an operative requirement at all times. All this also means detailed, intricate and extensive planning, assignment of resources and the time-frame within which all this has to be achieved. That is to say plans for logistic support must be made, vetted, and provisioned for well in time.

Parameters of a Logistics System

A logistics system evolves around identified strategy and goals set by a country. Strategy, in itself, is a differentiated component of defence policy, through which the means to attain national goals are organised and managed. And, the strategic posture of a nation is influenced by its economic capability to maintain and sustain such a posture. More often than not it is not possible to make major changes once strategical goals have been decided because a country's international commitments, its national ambitions and the course it is going to take for development are identified along side. National strategy always has long term plans and goals. A national strategy is, therefore, a very deliberate act of decision making at the highest level. Once the strategy has been formulated it becomes necessary to identify means both military and others to carry out that strategy. It is the strategy to be adopted that prescribes the "military structure", and those aspects like economy, industry, technology and will of the people that help in realising these goals. The structure of armed forces is crucial to their capacity to carry out given strategical and tactical tasks; especially important is ensuring proper logistics arrangements at each level of the 'force structure'.

The army (military), in turn, builds the financial and logistics planning and the manpower control processes around the development of operational plans that are in tune with the strategy. Such plans are based upon the requirements of the tactical situation to ensure delivery of the necessary troops and equipment at the time and place required. One cannot really economise except in choosing strategies (or tactics or methods) to achieve objectives. This means major economic and manpower investments in the military's preparedness. This implies that the economic and industrial capabilities of a nation limit the nature, type, and size of combat forces that can be created. At the same time, logistics capabilities limit the forces that can be trained, equipped, sustained and employed in combat operations. Thus, it is obvious that economic-logistic factors determine the limits of strategy. Again, the availability of resources, both currently and in the future, will govern the thought process. External and domestic factors are also reflected in the national strategies and the defence policies. Major international events like the break up of the Soviet Union or the liberation of Bangladesh (erstwhile East Pakistan) can bring in a situation which demands rethinking in terms of type, reliability and maintainability of the equipment in the hands of troops and redeployment of the military. Such major upheavals always mean some kind of a shift where national strategy, particularly with reference to logistics support to the military in carrying out a national strategy, is concerned. Further, logistics considerations inevitably will limit the size of the forces that we can sustain and maintain, and this factor will influence the revised strategy. More so, when we have to maintain troops in all kinds of environments and terrain throughout the length and breadth of our country. All this means a long-term perspective planning taking all the relevant factors into account. Where as the logistician is concerned, he needs to be informed about the specific requirements pertaining to a particular troop's deployment and to make an assessment of the resources required -as soon as possible.

At this point it is valid to discuss certain important inputs that go into providing a platform for providing means of achieving a 'national strategy' which in itself is governed by elements considered to be relevant by the government. These factors can be discussed under two compartments. First are the general factors that apply to the given situation. Second are those factors that relate specifically to the military aspects of the situation. That is, the national goals of defence can only be met if all the elements combine to move in the prescribed direction. General factors will include the political climate in the region, a country's economic potential, its industrial and technological base and the kind of military it can afford to have. The national strategy will definitely be governed by the geographic situation and location of a country and what role it envisages for itself. The approach of neighbouring countries towards regional affairs and a country's perception with regard to that approach will dictate what kind of military preparedness is essential for safeguarding own interests. The threat perception and the capability of our potential adversary will also influence decision making with regards to the strategy the country should adopt. Thereafter, the political masters of the day can decide as to what the country needs in terms of national defence and what it can afford. It will channelise the scientists to think either of going ahead on their own or to go in for import and/ or reverse engineering. This may necessitate even collaboration with certain other countries. It is rare to find a nation that has both the raw materials and the manufacturing capacity to produce all types of modern weapons and communication equipments required by the military. We have the capability to make small arms, light vehicles, and major components for tanks, artillery guns, etc. But we neither have the resources, or technical know-how to take on more intricate systems and sub systems in order to be totally self-reliant. In any case such an exercise which envisages total self-reliance cannot be cost-effective unless customers (among friendly countries) are identified who will be willing to buy our products. Alternatively, as is done in the developed countries we could go in for 'joint ventures'. There should be no inhibition in taking recourse to such an approach. Even more advanced countries depend on friendly powers to jointly produce their military needs.

For making a realistic assessment the logistician has to consider certain important factors. Broadly these factors are discussed below. All these factors are interrelated, and the list is by no means exhaustive.

Strategy

This is the most important aspect where military's deployment of resources is concerned. Technology—available and in the realm of the future—and economic situation of the country will eventually decide on the size of the support element of the military. There is really no point in formulating a national strategy if the same is not economically viable. Inter-relationship among strategy, tactics and logistics ultimately decides the course of action to be taken in a contingency. Anticipated action must blend the strategy, tactics and logistics factors. What it means is that the integration of strategic, logistic and tactical planning by those working in close physical proximity to each other is an essential factor in increasing the flexibility of a command. If either the strategic or logistics group has to guess at the requirements of the other, or delay decision until facts can be ascertained, flexibility is reduced. A sound logistics disposition and plan will support several strategic plans and a large number of tactical plans. The only operative clause is that any operational plan has to be logistically supportable. For not only are strategic and tactical plans limited by the feasibility of logistics support, logistic plans themselves are subject to the capability of the national economy because strategic reserves mean that much locked up capital. Since, it is often not possible to count on previously made strategic plans, it is necessary to be prepared to support any of a number of different plans or decisions within the means that are available or can be created. This means that the constraints in the form of funds, technology, non-availability of resources will always demand an approach that takes into account a large number of possibilities and the options available to the commander.

Strategic flexibility and mobility can only be based on a strong logistics foundation. Therefore, once strategic aims and goals are identified, the next logical step would be to assess the technological capability and economic viability. We should identify the mismatch between the two, assess the resource requirement, both in terms of money and material, and find ways and means of meeting those requirements that provide the right kind of national defence. But, strategy, technology, and economy are not three independent "considerations" to be assigned weights in a mathematical model. These are interdependent elements of the same requirement. Certain well known economists have defined that " Strategies are ways of using budgets or resources to achieve military objectives".

Technology

It defines the possible strategies. The economic problem is to choose that strategy, including equipment and everything else necessary to implement it. Therefore, one can say that strategy and cost are as interdependent as the front and rear sights of a rifle. The impact of technology on warfare is as old as the Iron Age. Throughout history the nature of warfare (and hence the relations between the states) has changed as a direct result of some new invention and its application to warfare. But to obtain real dividends from technology, particularly a new technology, it should not be used only to improve the existing operating systems; rather it should be used for seeking new ways of applying the military force in which technology can be used to its full advantage. There is a need to guard against the thought process that is dominated by a quantity-quality trade-off. New technology should be viewed as releasing the constraints upon present practices rather than, as we do now, being constrained by them. This is so because those tasked to exploit a new technology are often composed of elements committed by tradition and instinct to preserving their expertise in familiar, experience-proven areas. Rewards from new technology will not come from one to one substitutions of better weapons for old ones or in shoring up obsolescent doctrines (which should alter as the underlying technology fades), but in recognising the implications of technology for doctrine and organisation.

Economic Factors

These influence the objectives of strategy and they are intermixed among themselves and with strategic factors in a complex regenerative manner. In an equally interwoven manner economic-logistic factors influence the nature of the strategy to be employed. The issue involved here is that strategic goals of the country have to be influenced by the capacity of the national economy to absorb military requirements as the link between the war front and the home front. The logistic process is at once the military element in the nation's economy and the economic element in its military operation.

Tactics

Primarily, tactics are worked out to achieve the prescribed strategic objectives as assigned to the armed forces by the government. Tactics involve everyone, from the lowest level of the soldier to the generals. Tactical considerations define the logistical needs and also identify the material support required by the troops to effectively meet their goals. More often than not, tactics undergo a change as and when a new technology emerges or some lessons get learnt in some theatre of operations. We will not discuss tactics in this treatise except for stipulating that every tactical analysis and requirement of resources emerging therefrom leads to an examination of the logistical supportability of such a plan. After all the fuming and fretting, every commander eventually modulates his tactical plan in consonance with the organisational capability to back him up. There are enough historical records pertaining to every campaign where tactical plans, unless modified due to constraints of material support, ied to failure.

Information

Information with regard to defence needs vis-a-vis the capabilities of our adversary and our own resources to deter any threat forms an important input for a right decision to be arrived at. The strategy, the information with regards to logistics requirements of the prescribed military strength coupled with the necessary information on where, how and when can lead to arriving at the precise requirement levels. The controlling agencies could thereafter have the details with regard to the capability of each source for meeting these identified military needs. With the information thus available resource allocations can now be made to the industry wherein the peacetime requirements can be met. Concerned industrial sector(s) could also be given an indication as to likely operational requirements and the `lead-time' that would be given for rearranging their production cycles. In this, an important objective of the government policy should be the maintenance of up-to-date, modernised industrial production facilities, both in the defence and the civil sectors. The general efficiency of production (productivity) has a vital bearing on the amount of capacity normally devoted to civilian needs that could be converted to military production on mobilisation. The ideal situation for an Army Logistics Organisation would be to improve operations to the point that the primary constraint is the market and the level of availability of goods and services without undue strain on the community at large. This would mean that the organisational information system has to be fully effective and able to meet all the demands placed on it. It should be capable of giving sufficient and advanced warning to the suppliers to make their products available for military use on being given information to this effect. In order to avoid sudden hardships to other users of the products a specific time plan can be decided mutually for the production to shift purely towards military's requirements. In other words, the system as a whole would have excess capacity. All the customer demands would be met completely and on time during normal times. The excess capacity could be translated into a war time surge capability.

Intelligence

Logistics intelligence is what tells us what we should do to ensure that suitable, adequate and economical means are employed to safeguard the country's interests at all times and we are not caught off balance as it happened in 1962. This will also guide us in carrying out 'forward planning'. Timely intelligence will allow us to take the necessary steps for creating logistical resources well in time. This would also imply information with regard to technological advancements in various defence related fields being made elsewhere in the world, and ascertain our capability to adopt the latest technology that is in consonance with national goals. Information with regard to the capability of leading industrial houses of the country for technological upgrades to be available for defence employment when needed, will also be an important input because too much time is usually wasted in identifying and locating potential sources of supply for a very wide range of our needs.

Communications

This term implies not only information, but also the entire rail, road and air network we require to provide logistical support to the frontline soldier. An improved communication network gives us that much flexibility in moving forward stores and reserves well in time. Thus giving a commander means and an appropriate capability to carry out his designated task. However this cannot be done in isolation. This has to be a national effort. A national effort that will encompass the economic capability, its industrial strength and willingness of its people to pay the price such an endeavour demands. Of course, an improved rail and road network will naturally improve the economy of the region wherever such developments are undertaken. Developments that have taken place in our Northeast region after the Chinese debacle could conclusively prove the point that is being made here. Perhaps, the region would have remained isolated, inadequate and totally underdeveloped if defence needs had not arisen after 1962. Therefore, the country logically should not really feel a pinch if a rail, road or any communication network is taken up for development because of defence needs.

Elements of a Logistics System

All activities that go in preparing a logistics plan base themselves on availability of hard facts and information dealing with industrial, economic, and managerial resources. At the highest level programming is the heart of mobilising various segments of the system that operates to provide means of resources because a country's logistics capability is directly related to the capacity of its economy and industry to absorb military needs. For this to be correctly analysed a large number of inputs will be required. The thought process and activities that go to prepare a logistics plan are:

(a) **Organisation and provision of the basic concept of a plan**. This will be done at the highest level of decision making where an outline is defined based on the overall national strategy and goals, relations within the neighbouring nations, country's perception of its international standing, tactical situation, material capability (both existing and that which can be created) and resource availability. The plan will have both long term and short term goals. Based on this plan defence needs will be identified and action initiated to fill the gaps in equipment profile. Determining logistics requirements will involve use of planning techniques, forms and procedures. Planning factors will encompass basic usage data, initial correction factors and current experience. If the basic data is either outdated or a change of technology and development of communications makes it important to take note of the new scenario these aspects be also incorporated in order to arrive at a more realistic figure. For example, broadening of the railway gauge in the Southern and Eastern sectors mean that the carrying capacity has increased. This also means that the time and effort involved in carrying out trans shipments from meter gauge to broad gauge has also severely curtailed. During detailed analysis one could come to the conclusion that with this enhanced capability we could cut down on the range and depth of material held in those areas where communications have improved. That is, if we were earlier holding reserves of rations, fuel and ammunition etc,. for thirty days in the forward areas because of the time it took for stores to fetch up now we could cut it down to fifteen days because of faster movement of stores. The resultant centralisation could mean a very big saving in terms of money. A big pitfall is the factor of 'war wastages reserves' (WWR). The quantum, types and nature of these reserves are based on past experience. But in a large number of cases these either go back to the Second World War, or no realistic data exists for making an accurate estimate. With technology having gained momentum and equipment becoming obsolete in a much shorter time frame there is an urgent requirement to have a detailed examination of the subject. With improved communication systems, better training aids, more durable materials and higher education levels of the troops our data is certainly outdated and therefore not cost-effective. A serious and urgent re-look at the planning stage itself can help in bringing in realism and therefore, a more realistic picture of what exactly is required and where.

(b) **Modifications due to operational reasons**. The potential enemy's capability and his limitations, effect of time and space, weather and climate in the area of likely operation will have a direct bearing on the type, duration for which to be held and nature of reserves to be held. Special treatment of the equipment depending on these factors will be an important consideration. Any other major factor peculiar to the area will also have to be taken into account. The main equipment, accessories and other items and the time frame in which the troops can be trained and equipped to counter enemy capability in the area will be an important factor. Strategic dispersion of all warlike stores and ammunition is of vital importance. As WWR comprises a large volume of stores, ammunition and equipment, the stocking policy will be designed to achieve adequate dispersion as well as availability of reserves in close proximity to the fighting forces.

(c) **Evaluate limitations**. This could be the most critical factor that will give a realistic picture of the whole situation governing logistics. Above aspects do place restrictions and limitations on the strategic and tactical plans. It will pay dividends to realistically ascertain their impact on the degree of 'freedom of action' where equipping our military is concerned. There may be a requirement for modification of the strategic and tactical plans both in time and scope in order to ensure that our national strategical aims and goals are economically viable.

Concept of a Military Need

Hypothetically, the life of a military item begins as a gleam in someone's eyes. It was the invention of the 'stirrup' which made cavalry a more deciding factor than the infantry was. The idea for a new item not in the military inventory and not available commercially can be conceived in several ways:

(a) A soldier might conceive an idea on a battlefield, may be sitting on a lonely picket in the forward areas facing the enemy, as the enemy confronts him, by the elements, and a particular situation.

(b) An engineer in an industrial might conceive a new substance that, besides meeting a civil need, is also suitable for military application.

(c) A technician in one of our workshops might think of a special military use for a newly developed material.

(d) An officer in the training command (ARTRAC) might see the need for a special new weapon or piece of equipment to cope with a probable enemy capability of the future.

(e) An officer or any individual might get some idea from reading about a later date operation. He may even get the idea from some promotional literature sent by or advertised by the international arms manufacturers.

Very often all the above aspects telescope into a single perception about what could enhance our operational capability. This concept of the need for an item is thereafter evaluated with a view to validate battlefield requirements for the same and estimate the technological capability of either producing it, or an effort is made to locate it in the 'global market'. The catch is that the equipment that may be finally chosen may not be available in adequate numbers with an international manufacture. In such a situation the requirement could also be handed over to the research and development agency to see if it could be feasible to proceed with in-house development. This is a crucial phase of the logistics effort because it involves far too many aspects of the national resources, in the form of technicians, material, funds and all that goes into providing the latest to the troops. Competition with the enemy is both acute and chronic in the race to develop and produce or procure superior weapons and equipment. In today's technology driven world equipment can become outdated even before it is introduced fully into service. To guard against such a situation we have to ensure speed in decision making at various levels. Research and development requires a great deal of planning a multitude of decisions. A large amount of money and expertise in a diverse range of discipline will have to be pooled. Depending on the nature and the type of system or subsystems, a very big time frame would be required.

Introduction of New Equipment

To replace in-service equipment or to introduce something new, as a result of a shift in tactical doctrine, is often a time consuming process in view of the larger numbers involved. The need for prior trials and the need for induction into user formations that are widely dispersed means creation of maintenance facilities and provision of logistics support. This means coordination between various agencies. This would also mean changing the training system and maintenance policy and above all, bringing in a change in the habit of the troops who are used to particular drills and battle procedure vis a vis the equipment already in their hands. Therefore, the actual process of introduction of the new equipment needs to be carefully planned and executed if the changeover is to be smooth and without any prejudice to the operational preparedness of the services. Briefly, the procedure for equipment induction in respect of the army is discussed herein. Other services largely follow the same pattern.

Phases of Equipment Introduction

The lead-time required for introduction of new equipment may run into several years. Various activities involved in this exercise can be broken down into the following phases:

- **Long Term Concept Formulation**: Weapon and equipment policies are conceived in a long-term perspective of 15 years or more. They are governed by the 'threat perception' in the context of international environment, strategy of neighbouring nations and of nations interested militarily in the region. The "general staff" carries out this activity.
- **General Staff Policy Statements (GSPS)**: Based on the long-term concepts formulated the requirements of equipment up to 15 years are determined and enunciated through GSPS. These may cover a long time span of 10 to 15 years or a short span of Five to Six years. A pre-planned extension of the 'short term

policy' or advancement of the long-term bridges the interim period of 5 to 10 years if the development based on the latter policy is achieved ahead of schedule. GSPS are either of general application or relate to a particular range of weapons and equipment. Various agencies, apart from the general staff, such as R&D, the users, and the Department of Defence Production are involved. Note: Neither the maintenance agencies nor inventory holders are taken into confidence. This often leads to a situation where neither the repair nor spare capacity gets created. This denies the user adequate and appropriate exploitation of the newly introduced equipment.

- **General Staff Qualitative Requirements (GSQR)**: The GSQR is meant to reflect the user's needs to the development authorities. This is usually prepared in a draft form after consultations with the design authorities. The GSQR brings into focus inputs from different sources, such as the general staff concept, the user's need, the scientist's ideas and information from international and commercial sources with regard to a particular type of equipment. The focus of this information produces the rough image of the requirement, which is then circulated to all concerned in the form of rough QR. A GSQR lays down in broad outline the likely deployment of the equipment, physical, technical and functional characteristics, the number of prototypes required for trials, the probable date by which these should be made available and the necessary service requirements.

  On the issue of the draft GSQR the design authorities would carry out a feasibility study to assess whether the requirement of the forces can be met from within the resources available, both in the form of technology and industrial capability. At this stage, technical and financial feasibility, development and production, and economic viability of such production are considered. If indigenous development is ruled out, either because of inadequate technical 'know how', or acceptable time frame in which the end product is required, foreign collaboration for development and/or production is sought. If indigenous production is considered to be economically not viable, or urgency due to operational reasons demands quicker action, efforts are made to have the requirements imported.

The agencies involved are Master General of Ordnance, HQs commands, users, the R&D organisation and the Ministry of Finance, and others directly concerned with a particular type of equipment.

- **Design and Development**:The Defence R&D Organisation (DRDO) may undertake the development of an approved GSQR. During the course of their activities the end product may undergo changes, often beyond recognition. A study of the case histories of a number of development projects will show that very often the original factors in initiating the proposal are lost sight of and the development projects proceed as if these were the exercises that are ends in themselves. This happens for two reasons. Firstly, the users tend to make changes because some new technological breakthrough becomes known from some journal and they want to have the design upgraded. Secondly, the scientists, the factory tasked to bring out the prototype and DRDO in their enthusiasm and inexperience accept these demands of the user in the fond hope of giving him an ideal product. In following the trait of looking for a better and better end product at the behest of the users we reach nowhere and the forces remain equipped with equipment becoming more and more obsolescent. Classical is the case of MBT ARJUN. The practical necessity of striking the best available compromise between the time factor involved in development and production, the urgency of user's need, and the limitations imposed by the country's scientific, technical, production and economic facture has been ignored in the instant case. Consequently, besides the project going heavily over budget there is no end in sight for anyone to announce a date when the tank will eventually see light of the day. Main reason for this delay is our inborn tendency, particularly among the military men, bureaucrats and politicians in charge of a particular project, to clamour for more and more sophisticated growing technology whether the country has the capability to absorb the same or not. There are many such examples from the past where the military, particularly the Airforce, has been the subject matter of a number of Public Accounts Committee reports, mostly relating to the 'AVRO' project. Once the parameters are decided these be sealed and not altered at all. There may not be any obvious necessity to incorporate the latest technological advances into a system under development. Changes, if any, be made only when a particular model of the end product has been put through a thorough user's trial. And that too in subsequent models.

Entrusting to a Public Sector Undertaking. This is usually done when there is a development contract, as has been the case with our communication systems, or aviation spares. The task could also be undertaken by a Public Sector Undertaking on a proprietary basis. In the recent past, however, some private sector companies have also been entrusted with developmental projects on a proprietary basis.

- **Development Stage**: After making a 'feasibility study' and making its report the agency concerned goes ahead with the task of making an "initial development model" which is an experimental model broadly representing the physical dimensions and qualitative performance of the equipment. At this stage an assessment is made to see whether the concept of the employment of the equipment has been realised. Adequacy or otherwise of the existing maintenance methods, requirement of accessories, any additional training requirements and the impact of the equipment on the existing systems are also studied.

Based on the analysis of the initial model, a final development model is brought out where all the changes, alterations and any new feature later brought out by the user are incorporated.

- **Evaluation and Trials**: Technical evaluations are undertaken by the DRDO, or a 'Special Development Organisation' is formed for the purpose. In the laboratory or the development workshops the equipment or its model is subjected to functional and environmental testing to judge its efficiency in relation to the given QRs or the technical specifications. Thereafter, the equipment is subjected to 'user trials' by the nominated units. This trial basically ensures the following:

  (a) Performance of the equipment meets the QRs as sponsored by the users when deployed under service conditions.

  (b) Repair, maintenance, and defect-proneness aspects are thoroughly studied and evaluated.

  (c) Requirements of modifications or changes in the design are evaluated to meet the users and maintaining agency's needs.

  (c) Details of instructions on packing, preservation, transportation and user maintenance are worked out, and their adequacy and suitability established.

  (d) Maintenance scales of spares, Overhaul scales, EMER (Electrical and Mechanical Engineer's Regulations) are published in due course of time. At the beginning of a 'life cycle' of equipment, these scales are generally tentative, unless the newly introduced equipment is an improved version of something already in service. In all other cases the scales are often subjected to frequent changes and revisions thereby leading to defective planning of the spare part planning.

More often than not, there is an overlapping of various activities listed above. Although, the prescribed time frames are hardly ever met, the supplying agency in its hurry to grab a contract is generally not averse to cutting corners. This leads to situations where neither the maintaining agency nor the inventory managers get to know the full details of their responsibility. Besides causing avoidable confusion it creates an environment of uncertainty. To overcome the situation, of late, another feature has been introduced where "maintainability trials" are carried out. This helps in:

(a) *Assessing the life span of different critical components and sub-systems.* In our provision system, after five years of introduction of equipment the spare part planning is based on an actual consumption pattern. Since the equipment does not get fully exploited due to restrictions on its usage and very few training exercises, the consumption pattern does not get established in a realistic manner. This assessment, therefore, is expected to remove the existing lacunae and create a database for a more accurate estimate of spare parts requirement.

(b) *Establish common defect-proneness.* A weapon or a communication system has any number of sub-systems and assemblies. Some of them are apt to malfunction under different climatic or usage conditions. By carrying out such an exercise we can establish a specific pattern with regards to the situations that lead to malfunctioning of the equipment in a particular situation. If required, introduce modifications to enhance serviceability and reliability.

(c) *Studying stripping and assembly of components, sub-assemblies, etc.* In the case of weapon systems in particular, there is a need for periodic "strip inspections". This entails change of rings, oils, and certain other parts. By carrying out this kind of study the periodicity etc., are established for the more important maintenance tasks.

(d) *Working out the requirements of spares, special maintenance tools, and special test equipment*. Very often repair effort is impaired due to non-availability of certain spares, tools and test equipment. This usually happens when the original estimates are grossly wrong and fail to take into account various factors like quality of user's training, quality of the finished product and the like. Maintainability trials help in making more realistic estimates.

(e) *Finalisation and introduction of repairs and reclamation techniques, light, field and base repair efforts*. All these activities take considerable time, money and effort to establish. If undertaken concurrently the facilities come up alongside the introduction of the main equipment. Evaluation of special technical requirements is also carried out to ascertain if any additional effort is required in the form of either training or usage drills.

The above method of evaluation has several advantages. It confirms and affirms whether the equipment measures up to prescribed technical specifications or not, at the same time identifying the areas which may need modifications, changes, or alterations. Besides defining the periodicity of overhaul it also helps in firming up requirements of spare parts by assisting in drawing of realistic "Initial Stocking Guide" (ISG), and maintenance scales. With all these inputs one can work out the life cycle cost of the equipment much before a wider introduction in service is undertaken. The method also helps in overcoming initial problems and considerations, like where to establish 'base repair' facility, where to provide spares backing, who should ensure smooth transition from the equipment currently in use.

Another factor is that all the trials are generally rushed under 'somewhat' controlled conditions and by troops who have other commitments. Therefore, when the equipment gets into the hands of users in different terrains, the level of expertise and skills of the average soldier throws up a fresh set of defects and problems. These defects generally need modification kits, setting up of facilities and yet another cycle of procurement and supply with its inherent problems, and inadequacies. The biggest problem is that of the manufacturing agency wanting to cut costs by compromising on the quality of the raw material (s) required for various sub- systems. Even in the case of sensitive ammunition items the basic tensile strength etc, is compromised on the premise of "non availability" of the material of approved specifications and urgency of meeting commitments. The user is hardly ever consulted while these compromises are made. Generally, these reduced quality specifications surface when there are 'premature' failures and accidents. That such an approach leads to lowering of confidence of the user in equipment and ammunition is an aspect that needs monitoring at the highest level of command and logistics planning. What will help is if it is made mandatory for the manufacturers, be they in the 'public sector', or 'private sector,' to take formal approval of the indenting agency before changing any basic QR. The end user must be consulted before such alterations are agreed to and are allowed to be carried out.

Introduction into Service

After going through the drills given above the placement of firm orders bring in a stage where the equipment is introduced in service under the aegis of the MGO. It is hardly ever a smooth affair. Unless the user is clearly and well defined as in the case of tanks or guns there will normally be pulls and pressures, often on grounds other than merits of the case. This, sometimes leads to cancellations of "release orders", change in destinations and avoidable misplacements of costly stores. There have been situations where the 'generator' for a particular communication system has landed at location A and the equipment proper has gone to location B. This happened because the issuing depot staff had already loaded the generator in the wagon as per the original release when fresh orders arrived. The lower level staff either ignored revised instructions or were unaware of change due to a lack of communication. The problem is further compounded when one half goes to the east and the other half to north.

Economy in Logistics

Economic efficiency in the military services, as in households and business firms, is the responsibility of everyone, not of some specialised official or group alone. Sometimes it requires nothing more than straight hard thinking about objectives, and the arraying and intuitive judging of alternative means of accomplishing the objectives using an appropriate economic criteria. Frequently however, special skills are necessary, or are highly desirable, to make the requisite analysis. It will entail expert military judgment of scientists, engineers, economists and other social-scientists. It is, in other words, the kind of function that is likely to be performed better by specialists. In these days of "resource crunch" and shifting priorities one sees too much effort being made to reduce inventory and to cut inventory-carrying cost. But we must remember that, "If the logistic system in question is not to be hopelessly fragile and liable to catastrophic breakdown," as one observer of military affairs writes, "if it is to function under changing circumstances and be capable of switching from one objective to the next. If, in short, it is to be capable of coping with the uncertainty that is the result of enemy action and, as such, inherent in war- a certain amount of redundancy, slack, and waste must not only be tolerated but deliberately built in.[1]" A number of studies after ' Operation Desert Storm " have concluded that if the USA and the military of other participating countries had gone in for the traditional methods of inventory management followed by the industry it would have taken them " two years " to go to the aid of Kuwait.[1] Because that is the lead-time their industry needs to meet defence requirements. In our case it is anywhere from 36 to 60 months for the public sector units (PSUs), and in the case of 'welfare oriented' governmental agencies it could really be open ended. Therefore, the operational commanders will have to be very careful when trying to enhance their capability by adopting these "industrial management techniques".

Logistics in the System Life Cycle

Logistics in the context of the system life cycle[2] involves planning, analysis and design, testing, production, distribution and the sustaining support of a system throughout the consumer usage period.[2] Given a specific need, the system operational characteristics mission profiles, deployment, utilisation, effectiveness figures of merit, maintenance constraints and environmental requirements are defined. Effectiveness figures of merit may include factors like cost effectiveness, availability, dependability, reliability, maintainability, and so on. Using this information, the system maintenance concept is defined. Operational requirements and the maintenance concept are the basic determinants of logistics support resources. In this the primary concern is to ensure that the equipment sustains highest level of reliability throughout its life cycle so that the combat soldier, sailor or airman retains his faith and confidence in the product. This, in turn, ensures his success in war. This presupposes that the weapon and equipment provided to him are the best that the current technology can provide within the budgetary constraints. This means ensuring that the man on the front has the latest tools of his trade. Equipping of a fighting machine, therefore, is a dynamic process, and the vision of the policy makers and planners in the armed forces will ultimately decide the efficiency of the weapons and equipment with which a soldier goes to war, as and when it comes.

Notes and References

1. Martin Van Creveld, *Technology and War*, pp. 316-317.
2. Classically the term means womb to tomb or the stages of conception of the idea for an equipment, identification of users, its development, production, operation including repair and maintenance, and its final disposal.

3

Principles of Logistics

"Administration in War" pamphlet issued to the British military units (our pamphlet also says the same) lists certain principles of Logistics support. These are - Foresight, Economy, Flexibility, Simplicity and Cooperation. As simple as that. The Americans have nine principles. These are- Logistics Intelligence, Objective, Generative Logistics, Interdependence, Simplicity, Timeliness, Impetus, Cost-effectiveness and Security. The main problem is that, at times, commanders tend to function under the assumption that the traditional "principles of war" are equally applicable to logistics. This is not so. The correlation is, at best, only partial and governed by relative context of a particular situation. Not understanding the subtle nuances commanders in their anxiety to measure up to the operational plans, often and at their own peril, tend to ignore these principles that govern logistics. They tend to forget that a proper assessment of the logistics support element adds value to the sustainment and success of operations in several ways. It can bridge the gap between capability and requirement, particularly in the earlier and late stages of the deployment when the situation is, as yet, not very clear and the unknown elements of the operational side can become serious impediments for successful conclusion of the operation. By habitual use, or by results becoming evident in ignoring them, certain prescriptions may be said to have come to light which, with necessary reservations and qualifications, may be offered as tentative principles of logistics. Not as unalterable laws or maxims, but simply as general guides. To be responsive to operational needs of the troops there are certain basic principles that govern the making of resources available to the troops.

Americans, superbly confident, after the Second World War went into Vietnam either ignoring these principles or thinking that what worked against the Germans would do against the "Vietcong" as well. Consequently, they suffered heavily when their logistics either failed to respond or were found wanting when deployed in the marshy land against a clever and motivated enemy fighting a "guerrilla war". Over a period of time they realised that they had to relearn their logistics applications. (Logistics security, including the physical protection of the Logistics personnel, installations, facilities and equipment was one of the more critical aspects of the logistics effort in Vietnam. The tactical situation is not always evident or given consideration during the installation construction planning phase. Remember there are no secure rear areas.)[1] It is virtually an international phenomenon that the 'principles' that relate to logistics are so often violated in actual practice that it is now necessary to put them down as "Principles of Logistics". It is only appropriate that we examine those factors that should govern our thinking while planning, positioning, and carrying out the task of supporting combat troops. Certain facets on which a logistics policy is to be formulated also need to be discussed so that a common platform both for the combat and the combat support commander can be created. Basic characteristics of logistics—the factors that are required to successfully support the soldier on the battlefield—remain timeless. Primarily these are:-(a) Equivalence (b) Mobility (c)Momentum (d) Flexibility (e) Dispersion (f) Feasibility (g) Timeliness (h) Sustainability (j) Continuity (k) Accountability (l) Maintainability (m) Standardisation (n) Integrability and Interdependence (o) Communications (p) MIS . We will discuss these principles and their applicability in various operations.

Equivalence

When the author was part of the logistics services of a Corps headquarters the Heads of Services were not even shown, leave alone involved in developing the operational plans of the formation, primarily on the premise that logistics do not rate a predominant consideration for an operationally committed corps because the resources were supposed to be made available by the "higher headquarters" in any case. And in so far as the general staff was concerned we i.e. the Services were to be told when, where and if required. This kind of secrecy basically stems from the fact that the commanders who are not aware of the efforts involved in making their plans practical do not even want to know. Where they are concerned their attractively made 'sand model compatible' plans which have been accepted by equally (logistically) ignorant higher commanders need no consideration by their advisers who were there to make resources available for the plans. It took a lot of persuasion to be allowed a look at the operational plans of the formation. Not surprisingly (to my horror), it was found that plans for "drawing of ammunition" had been made without any consideration for the capability of the 'ammunition dumps' to handle their designated loads. Neither the manpower nor the infrastructure for the number of vehicles which were expected to draw their requirements from those locations during the operations was available nor had it been specifically catered for. In some cases, located over 12220 feet above sea level, the dumps would have needed a labour force duly acclimatised and sufficient space to handle the "anticipated number" of vehicles coming to draw their ammunition. The operational plan was not supportable in the given circumstances. Somebody should have realised that the particular ammunition dump did not even have adequate turning radius for the types of vehicles that were expected to come to collect their ammunition during operations. Nor, for that matter, it had reliable means of communications within the establishment or with the outside world. No plans even existed for provision of communication within the dump that was spread over 11 kms along the road. Ostensibly, it had been overlooked that logistics communications (particularly in the case of ammunition dumps) are normally over far greater distances, and carried out with far fewer communication assets, than those of the combat troops. Consequently, things can go wrong. It is difficult to fathom the reason for this apparent lack of faith in taking the Support Services into confidence while making the operational plans or while selecting site for important logistics element like an ammunition dump. It is always difficult to convince the "general staff" that an operational plan made and rehearsed without taking their 'Service Advisers' into confidence have a very limited chance of success. A clear and simple logistics concept which is in the minds of all those involved means that, when communications fail and other things go awry, individual staff officers and logistics unit commanders can take decisions in line with the general thrust of the overall plan. And this can only be done if the supporting agencies are involved in the whole exercise and they contribute professional expertise and advice while the plan is being developed. There is a need to plan tactics and logistics concurrently, otherwise considerable effort could get wasted when eventually it is found that the plan could not be supported for reasons other than tactical. This means that the logistician must become intimately familiar with the supported commander's plan. After careful analyses of the material needs he should provide inputs for appropriate placement of resources for various phases of an operation. He should have active and interactive dialogue with the operational commander where optimum capability to support various operational aspects are critically examined. It is required because the locus of operations of field-forces may change rapidly and there could be a break in communication for some time leading to changes in the support plan. For successful execution of any operational plan consistent support must exist no matter how great the change in nature of the field operations might be, and field operations hardly ever follow a predictable pattern. The real test of logistics support is that it is capable of absorbing most of the operational situations, be they predicted or unpredicted. All this can only be achieved if the expertise of the logisticians is utilised when operational plans are being made and logisticians are treated as equal partners in the whole exercise.

An equally important thing is that most of our Army doctrines have their roots in the campaigns of Second World War, which were characterised by definable fronts, serviced by reasonably dependable and well defined supply lines to the rear. In today's world, when technology has broken down the barriers, bringing both the combat and combat support soldiers within the enemy's air and missiles there is a definite need for the combat commander to involve his logisticians in all stages of planning. This will help our support systems to measure up to all kinds of predictable and unpredictable operational situations. By actively and constructively involving the logisticians in early stages of planning, the commanders will be able to have a professional assessment of their material needs. This will give enough time to the support services to work out the shortfalls and find ways and means of meeting operational requirements of commanders. It is time it was realised that war is a multi-faceted affair that cannot tolerate parochial intransigence; all the parts must work together or the results can be counter-productive. Strategy is to war what the plot is to play. Tactics is represented by the role of the players. Logistics furnishes the stage management, accessories, and maintenance.[2] Granting equivalence to the three facets viz. Strategy, Tactics and Logistics of an efficient war machine is, therefore, an important ingredient.

Mobility

'Mobility is the true test of a supply system'

B. Liddell Hart, *Thoughts on War*

Sometime back a 'sand model' exercise where "Armour" formations were deployed in the desert was conducted in a training establishment. It was an 'Army level' exercise and things were generally moving as per the 'Whites' when due to some hectic and heated discussions regarding the operational situation urgent need arose for switching the huge "tanks concentrations" on to an all together different axis in order to counter enemy action. The moderator coolly lifted his 'tank models' from one end of the desert and put them about 150 miles away and resumed the discussion vis-a-vis the changed battlefield scenario where the armour had now been re-deployed to neutralise the enemy. The author got up to ask him about those large quantities of ammunition and fuel that had been collected at the previous location. Those could not have fetched up on their "soft" vehicles to the revised theater of operations as no "tracked" vehicles had been made available to the Support Services for providing them with commensurate mobility. Therefore, it was felt that the commander was not really ready for the next phase of his battle. It took some time for this to sink in that mobile forces require flexible logistics systems and mobility of the field force will be greatly limited to the extent that the logistics system constrains it.

Logistics activities themselves must be mobile to permit rapid displacement for continuous replenishment of forces. This is necessary in any situation, whether in supporting limited or extended offensive pursuit, defensive or retrograde operations. Not an easy thing to do, mobility of logistics units must be put to test and where required modified while conducting tactical training and exercises. This is particularly important when mobile formations are exercised and are being fine-tuned for their operational roles. Location, identification and even participation by local agencies who could contribute during an operation should also be undertaken alongside so that all the likely participants during a likely operation are made familiar with their roles. There is no simple answer to the logistics problems in the deserts, as perhaps was, and wrongly being projected in the 'sand-model exercise' mentioned herein. One must, however, also accept the fact that it is really difficult to simulate and practice the movement and handling of large quantities of ammunition and fuel when the same have not been actually transported and consumed. A serious discussion did take place on this aspect of how to provide mobility to the logistics units so that the operational plan was not hindered.

Here it will be relevant to quote from a US army exercise where a computer simulated exercise was conducted by their corps with mission of conducting a deep attack to seize an operationally critical objective. It was established that the Combat Service Support (CSS) had as much relevance as the Main Battle Area (MBA). It was accepted that CSS also has to go through the intensity of close combat as well because logistic units also have to fight distance, time, interdiction and confusion. They have to ensure that forward-deployed tactical units are capable of continuing close and deep attack battles. It was experienced during the exercise that to support a deep attack, the Corps 'rear area' requires centralised command and a direction on the same order as a division sector. Other lessons were:

(a) CSS units cannot and should not expect substantial combat forces to be dedicated to rear area defence. They must be able to stand up against an encroaching enemy force and defeat it without requiring the diversion of combat strength from the main battle. It must be manned and equipped to execute the dual missions of support and defence.

(b) CSS units must be placed forward to provide support to the deep attack force prior to crossing the forward line of own troops.

(c) A sub-set of the larger problem of command and control in the corps area is movement control and terrain-management. The present doctrinal (of US Army) approach to movements addresses the control of main supply routes (MSR) must be integrated into the overall support and defence plan of the corps area.

(d) At the corps level over the long term, rail is the preferred method of support due to the magnitude of the support problem.(In our case it will have to be a mix of rail and road because in certain areas the rail hasn't reached so far).

(e) Although aerial supply is most effective, aircraft survivability and lift capability becomes a major limitation. Aircraft and helicopter alone are not sufficient to move the thousands of tons ammunition and hundreds of liters of fuel needed to support large scale operations.

(There is a need for such simulations for various terrains and all types of operations in our environment also so that a more realistic framework of logistics support can be developed and practised in order to provide mobility to the support services commensurate with their designated role and task).

Simulations, also, clearly showed that once the shooting starts, there is very little distinction between tactics and logistics. Therefore, unless there is close and intimate relationship between the man doing the fighting and man providing support to him there could be serious trouble. Today there really are no safe rear areas in war and both will be exposed to same or similar dangers. Logisticians have to be tacticians and the tacticians must be logisticians. Both must understand the interdependence of all functions within the mobile battle sector. Regardless of all the planning, provisioning, bold decisions on the 'sand models' in the final analysis it is the willingness to over-ride plans, to improvise and take risks that determine the outcome of a real battle. It is wrong to assess strength only by the number of tanks and guns that can be deployed. These are just a part of their strength. In a real war, without a ready, efficient and continuous supply of ammunition, fuel, rations and replacement equipment the fighting force will soon lose its teeth. There are any number of instances where apparently successful campaigns got stalled for want of commensurate mobility for the material requirements of commanders conducting operations. A commander must always be conscious of the fact that a slow logistics system and inadequate manning levels militate against achieving the best potential with existing equipment. Only too often successful break-through of the enemy's lines have grounded to a halt because of lack of supplies necessary to sustain a continuing advance. The planners had obviously assumed that some how things would work out but they didn't. There is no harm in our learning lessons from history and from experiences of those who have gone through these situations. Taking those actions and situations as guidelines we could decide on our priorities with regard to the type and quantum of stores required to support mobile operations and assign appropriate type of transport so that load tables etc., could be worked out and rehearsed. For example, it is not difficult to see that a small percentage of tracked vehicles could surely help in carrying urgently needed items to tide over immediate developments in the ever-changing operational situation when the planned operations are in the deserts or Armour intensive.

How to Provide Mobility

Mobility is essential to support fast changing situation in any type of terrain or operation. In order to ensure that the material support is provided to the troops at all times we should be able to use all the means available in the country. The bulk stores should be moved by rail as far forward as possible , for this even if we have to lay rail network it should be done because the rail will also improve the quality of life of the locals as well in addition to providing cheaper mode of transport for military stores. In the North-East where a large deployment of troops has been in existence for a long time now rail either ends too far away from the logistics units or the gauge differs from the base depots. Thus the rail network is serving a very limited purpose. Similarly, it is only now we have undertaken conversion of meter gauge into broad gauge in certain other sectors. In order to ensure mobility of the material support such projects need to be hastened. From the rail-head, depending on the terrain heavy or light vehicles can be used taking supplies as close to the forward troops as close as possible. With the kind of "resource-crunch" based military that we have there could be a question of finding funds for an appropriately 'mobile material support' for the field force. A little resourcefulness, imagination, ingenuity and a will to try something different can help overcome the problem. (In Viet-Nam the famous "Ho-Chi-Minh" trail was used by the communist forces for conveying thousands of tonnes of ammunition and other stores through hilly terrains on bicycles when no other means were available. Ammunition had been specially packed in a manner where the fuse was carried on the handle, projectile along the front frame and the cartridge on the rear carrier. This made it possible for the troops to be constantly and consistently supplied with their requirements with limited engineering support as the bicycles hardly required any high-tech maintenance and repairs.) For every 'remote area' deployment of our troops we should also consider innovative ways of providing mobility to our support element. Only then the soldier will have some kind of confidence in his supply system. We have a fairly large number of obsolete and discarded tanks, APCs and certain other types of tracked vehicles that are usually sold off as steel scrap. One could have these vehicles suitably modified to provide commensurate mobility to the intimate material support columns in the deserts. In so far as the plains or the mountains are concerned we can try some of the combinations, like roadways, and "rope ways" being tried elsewhere in the world. The crucial aspect is that the task of ensuring mobility of fighting forces and their capability to sustain and retain momentum cannot conceivably be delinked from the over all operational plans.

The increasing sophistication and diversity of weapon systems has led to a more equipment intensive army. The tendency towards larger caliber in artillery weapons, the increased number of tanks to be expected on a battle-field is the need of the day. Consequently, the anticipated heavier expenditure of ammunition plus the ever present need to feed and clothe the soldier, have created a demand for a whole range of medium and high mobility vehicles for use by the logistics units. This has lead to a situation where one of the more perplexing issues facing the commanders and logisticians is the challenge of supporting forces in the field. What makes the task even more difficult is the aspect that we have such a wide range of equipment and the areas of operation that one cannot identify a common platform. Any solution for the deserts may not suit the plains or the mountainous terrain. In the case of the desert operations the tanks and self-propelled guns need to refuel while on the move in order to achieve higher rate of mobility. Tracked tankers can look after this need. Similarly, essential ammunition for the guns also needs to be carried on tracked vehicles so that immediate replenishment is available to the troops committed in the operation. Tracked tankers and ammunition carrying vehicles are yet to meet bureaucratic approval. So the commanders have to do with what is available or what can be made available within the realm of "ministerial sanctions". In the case of mountains, stores move by rail up to the "rail head". Thereafter, the road transport takes them to some intermediate location. After that it could be still lighter vehicles, mules, porters and what ever could meet the requirement. With so many transshipments, handling by different agencies, and depending of the packing of these stores we do have a serious problem of maintaining and providing mobility to our forces.

Another area of concern is inadequacy of field communication for the Service units. Since priority is given to the teeth arms the Services have to make do with the "left overs" or whatever can be conveniently made available. This is not a satisfactory arrangement, particularly in mobile operations when the situation is too fluid for any pre-planned replenishments to support an operational situation. The advent of more effective and smaller radios should permit appropriate communications to be provided at all levels. Convoy commanders and even individual drivers could be given these sets so that a change in operational situation from the time the vehicles had left their base can be communicated to them while they are on the move. Simple and easy to operate radios should be provided for the logistics vehicles. This would provide a means of re-routing supplies whilst on the move and there is a change in the operational situation since the vehicles had left their bases. Empty returning vehicles can also be directed to the correct dumps and re-supply points.

Role of Railways in Providing Mobility

With the invention of steam engine the military planners of that era quickly saw the potential of this means for conveyance of military stores to the area of operations. It had already been proved that horse-drawn carriages were both inefficient and inadequate for large-scale movement of ammunition, rations and other war-like stores to the areas where road network was either primitive or non existent. In fact when Hitler came to power- having realised the importance of the rail- he took special measures to incorporate the use of railway for communication of men and material. It was realised that given the technology of 1939, no less than 1,600 lorries were needed to equal the capacity of just one double-tracked railway line. What is more, their greater consumption of just about everything (fuel, personnel, spare parts, maintenance) in relation to payload meant that the railways retained its superiority at distances of over 200 miles. It followed that, while motorised columns were essential for operational and tactical purposes, their effect on strategy would be limited because strategic logistics invariably formed very heavy tonnage. Allies also built a lot of rail networks, particularly for the Burma front and those parts of South East Asia where they had to create logistics facilities against the Japanese. The mix of broad and narrow gauge caused a lot of problems when transshipment of heavy and unwieldy stores like bridging equipment and guns etc was involved because technology had not yet caught up to provide suitable material handling equipment. But it was felt that taking stores as close to the forward troops by rail and thereafter use of alternate transport reduced the turnaround time considerably. We have established our Base and Regional depots in locations where rail can take stores as far forward as possible in an economical and efficient manner. In certain areas the railway gauges are also under conversion.

We have problem of a different kind where use of railway for movement of stores is concerned. Unfortunately, and rather surprisingly, it has always been difficult to get railway wagons for movement of Military stores to forward areas from our depots. The provision of wagons to military has always been uncertain, unpredictable, rule bound where the so called 'day light hours' are concerned and ridden with all kinds of "booking restrictions" and other measures that do not add to the efficiency, either of the railways or that of the military. Now the railways, in an effort to generate profits, even discourage provision of any thing less than a 'rake'[13] to the services. Other than a bulk consignment of ammunition, petroleum products or certain 'ration items' being moved from the base depots to forward depots there usually is no reason for the Army to ask for the rake. Very often we have to wait too long for some body to provide wagons individually for different destinations, leading to a situation of uncertainty. To a logistician who is required to provide materiel support economically this does not seem to be the right way of contributing towards national defence.

Consequently, and expensively the operationally required stores mostly now move by road. No body seems to be realising that the use of road transport to carry logistic supplies over long distances is both inefficient and uneconomical compared to the railways when the load is not enough to make a rake and only a few wagons are needed. That, such close minded policies have an adverse effect on the mobility of the material support to our troops hasn't been registered in any decision taking quarter as yet. There is essentially a case for formulating broad government level guidelines in order to see how we achieve economy in national logistics systems, which provides adequate mobility. All that needs to be stressed upon is the fact that there is an urgent requirement for effective co-ordination in evolving production, transportation, and allocation policies and their implementation at the administrative decision-making echelons of defence ministry, other important ministries like Railways, Shipping, Surface Transport etc. The important thing is that the combat soldier must always be certain of material support for his operation in hand. Various methods for ensuring this have been tried by different armies of the world. A mix of rail, road, inland water transport, and air including helicopters has been found to be ideal for various operational situations. The Americans refined their forward support system in Viet-Nam where the terrain and enemies resourcefulness made them experiment with various combinations. They further improved upon it when they went to help Kuwait against Iraqi occupation. Things like 'closed loop support' and 'horizontal support' (discussed elsewhere in this narrative) were refined during these major operations. The underlying theme should always be that the tactical commander be allowed to concentrate on his primary mission. The commander should know that the required and designated logistics support would be available whenever and wherever required. Briefly, let us examine what others are doing to resolve the issue of material support to their mobile forces.

European Experiments in Providing Mobility

Mobility of the logistic support units commensurate with that of the supported troops has been causing anxiety in most armies of the world. Consequently, a number of European Armies, including the British, are carrying out field trials of, what they have termed as, the DROPs (De-mountable Rack Off-loading and Pickup) system of logistic supplies. The basis of DROPs is what is known as a ' flat track', a platform of 20 by 8 feet standard container size. This forms the cargo platform for a special truck, up to ten pallets or unit loads. The whole flat track is loaded and off-loaded on to the vehicle using an ingenious hydraulically powered handling system, which is normally an integral part of the truck. Flat racks can be easily loaded on to flatbed rail wagons and then transferred by the vehicle's own equipment on to the truck itself. Another concept being experimented with is the formation of a 'Composite Logistics Regiment'. The British during the 'Falkland war' tried this for the first time. A permanent Logistics regiment in the Allied Command Europe Mobile Forces has since been formed. The commander of this regiment is responsible for logistics supply up to Brigade level. Also, as a trial measure, a ' Composite Maintenance Group' has been created in the base or army area. This group controls all the Supply, Ordnance and Workshop units and the commander is responsible to the 'Chief Logistics Officer' at Army Head Quarters. Responding to the demands from the 'Corps Logistic Officer, logistic requirements are dispatched to the 'Corps Replenishment Points' and from there to the 'Exchange Points' in the divisional areas. The next move forward is what is known as 'Immediate Replenishment Group ' (IRG). The 'Brigade Logistic Regiments' establish these. As far as possible loaded vehicles move all the way through the system and are not unloaded until the final IRGs. From the IRG onwards depending on the terrain smaller or tracked vehicles are used from the units. Tanks and APCs can refuel directly from an IRG tanker wherein a system known as Forward Area Refueling Equipment (FARE) is used. A tank company can be refueled completely in less than 20 minutes. They have also experienced that refueling on the move gives an 'armoured unit commander' a logistics-edge, regardless of the distance to be traveled. His capability to sustain the fight is increased because the combat unit's organic fuelers arrive in new positions with their complete basic load of fuel. Exercises involving refueling is regularly conducted to enhance the tactical advantage of repositioning task forces.

The NATO System

In the American army 1st Cavalry Division have implemented what has been termed as Unit Level Logistics System (ULLS). In this, company/squadron level containers wherein all logistical needs including ammunition are packed as a single unit are used. These are delivered as far forward as possible. Americans have also designed CONEX (Container Exchange) on these lines. In this system something like the railway 'container service' is used for carrying forward all the material requirements of an "Army Aviation Unit". This was basically developed in Viet-nam where "pre-stocked containers were sent to various units providing a 60 days stock level of spare parts (some 95000 items) in 70 semi-trailers and 437 CONEX's. Now the NATO forces and the American forces are thinking in terms of a "containerised ammunition distribution system (CADs). It is a 20*8*8 foot steel slipping container. Packed at the factory the, the container can be handled safely and securely during transport. These can also be transported safely and directly by rail to the storage site. With the limitation of "ammunition storage space" that we have in our ammunition depots this might be a good idea to try out developing this kind of facility. This will serve dual purpose of providing temporary storage space as well as that of speedy loading and move of heavier ammunitions to their area of requirement. The benefits of using containers are:

- Increased efficiency at trans.-loading points. Instead of unloading individual packages pertaining to different types and nature of stores and then sending for transsshipment the container as a whole can be speedily shifted.

- Greater protection and security for the cargo. Since the container will be properly locked and sealed chances of pilferage en-route decline considerably.
- Track of "in transit" stores can be kept more easily. Due to very wide spread deployment of the troops, and distances from the base depots being large, it is not unusual for consignments to be missing in transit, often for months together. A railway consignment sent to Jabalpur may land up in Jamalpur causing all kinds of problems and inconveniences. When this happens the railways do express their helplessness but it does no good to the functional or operational efficiency of the troops.
- A smaller number of people required for unloading. Modern mechanical means of lifting and moving logistics supplies can be used for this purpose. This will speed up the process, and at the same time will require lesser number of men for the task.
- Optimum use of technology and capabilities available in the commercial sector. Even our troops on UN assignments can be more efficiently supported by dispatching containerised consignments in commercial ships and vessels.
- Greater potential for throughput as far as forward as the operational situation allows. Containers can be moved directly from the supplying agency to the end user.
- Container can easily be used as additional covered storage accommodation facility in the forward areas when the troops are likely to move elsewhere in due course of time.

Russian System

The (erstwhile) Soviet Army in Afghanistan had experimented with what was termed as 'Manoeuver Support Battalion' (MSB). Besides, all the conventional Service units the establishment also has an Engineer's Platoon with plants and dozers for clearing and maintenance of roads (very often the Services take a lot of time to catch up with the supported troops for want of appropriate road network or inadequate tracks. We could do with some Engineer effort, including bridging, integral to our 'Service Element'. In certain armies "combat engineers" and other complements like dozers, plants, etc are separate entities). Russians are understood to have combined in single units all of the major divisional assets for material support: units that receive, transport and supply ammunition, FOL, repair parts, food, clothing and other items. These units have been assigned increased evacuation, internal maintenance, and rear area security duties. They have, thus, united the disparate support elements under one commander. A major consideration in this has been the need to find more effective ways to supply task-organised, combined arms maneuver units assigned to operate on an axis separate from the main body. (This kind of situation more often than not prevails in our army where the Ordnance and the Workshops etc. are required to provide detachments for which they are neither designed nor equipped). The material support battalion commander is subordinated to the division's deputy commander for coordinating and controlling the division's logistic activities. The Bn HQ has several "general purpose" trucks, utility vehicles and communication vans. Mobile Bakery consisting of sufficient number of mobile field kitchens, some medium trucks and water trailers provides support to the combat troops. Transport Companies, and FOL Company, having medium cargo trucks with trailers and need based FOL trucks and trailers provides mobility to the support services where they can meet the troops requirement. There is a " Combined Department" that carries stocks of ammunition, spare parts, food, clothing items etc. It also has cargo fuel trucks and material handling equipment. Maintenance Platoon carries generators, utility and recovery vehicles and the Engineer's Platoon is equipped with dozers, earth moving equipment and stores required by the "Sappers". In addition to the above there is a 'general-purpose cargo transport company equipped with at least 60 medium and light cargo trucks.

Recommendations for an Indian Model

The thumb rule is that it is more important to have material "in place" than to have unequipped forces "in place" if we want to ensure tactical success. The key to rapid mobilisation is the availability of weapons and equipment. In our environment we can examine feasibility of a combination of models discussed above. In that, for our Deserts and Plains formations we can 'broadly' have the NATO system. The Army can enter into a contract with the "Container Corporation of India(CONCOR) convey stores from the base depots to the Rail/Road head of a formation. Army can also participate in the "own your wagon" scheme of the railways. This will eliminate the delays presently taking place due to non-availability of wagons for "piecemeal" dispatch of stores. Advantages of using containers have been discussed earlier. Induction of "containers" and own wagons will help in mobilisation of reserves. For Mountain and High Altitude areas the Russian model offers us suitable guidelines for restructuring our 'Logistics Support System', particularly where it consists of incorporating "Engineers' complement" and other essentials like FOL and ammunition. All the supporting Arms and Services can be placed directly under the deputy GOC of the division for effective monitoring of 'resource movement' in the theater. This will enhance mobility of vehicles carrying essential stores and medicines etc., particularly when the 'cross-hauling' of materials is taking place within the operational area. The activity (of building up stocks) should be confined to strategic logistics in the first instance because the enemy can easily discern motives from a logistics build up and take measures to deny the advantage. Strategic mobility[4] requires more than just lift assets. Infrastructure- such as railway sidings, marshaling yards, airfields, access roads, and loading facilities- at key installations and important base depots need to be kept fully functional. Border area petrol pumps, service stations etc. be fully exploited for this purpose.

It will be seen that everywhere the thrust has been towards total integration of logistics elements and commensurate mobility vis-a-vis the type of formation supported during a particular kind of operation. The "principle" is that the mobilisation of heavier ammunitions, fuel etc., should precede manpower mobilisation of troops. With the technology having improved the material handling equipment and capability of cranes etc. we can experiment with some of the measures discussed above that will provide enhanced mobility to the material support complement. Use of containers, combination of all possible means of transportation and bringing in some organisational changes that allow us a better and economical use of resources can be tried. Heavy demand on FOL will always remain, and on their ready availability will depend the mobility of troops. In more advanced armies the availability of light weight easily handled pipe lines and bulk refueling from powerful tankers and the use of transportable fuel pods have virtually replaced the large number of jerricans required in the forward areas. This gives greater mobility and faster reaction time to both the teeth and the tail. When mobilising for an impending major operation it will be appropriate to ensure speedy placement of resources in the theater.

Another medium for providing better mobility is the inland or water transport system-IWT. It was used quite effectively in Burma and later the Americans made full use of rivers and rafts in Vietnam. They even used barges as floating "gun platforms". Main limitation of this system is that it is seasonal due to flooding of rivers and is effective only where both the origin and destination are located on river banks.

For our purposes there should also be suitable roads at the bank sites for loading and unloading of stores. IWT can be particularly effective in maintaining our formations located in North-East.

All the available and potentially deployable mediums of transportation should be used as much as possible while carrying out peace time movement of stores and troops so that those military and civil personnel who handle these facilities are fully conversant with their war time roles. Exploitation of the existing road-transport system when moving stores that do not make a wagon -load is also essential in peace-time because besides allowing for compact packaging it avoids repeated handling as is required in case of rail-transport. That the chances of pilferage en-route (a perennial problem in the case of rail moved stores) decline is an added bonus. Containerised stores where platoon or company level requirements can be kept pre-packed and ready in the base or regional depots to be moved quickly to the area of combat troops deployment is apt to give great amount of confidence to the supported troops.

Flexibility

> It is no great matter to change tactical plans in a hurry and to send troops off in a new direction. But adjusting supply plans to the altered tactical schemes is far more difficult.
>
> General Walter Bedell Smith in ' Eisenhower's Six Great Decisions'.

From a logistician's standpoint, the military is a structure that depends both on flexibility and on rigidity. The doctrine is rigid- but only up to a point. Past this point, it becomes extremely flexible. It first lays guidelines, but it then demands innovation, adjustment and foresight. To the logistician, who is paid to look at reality, in order to be anticipatory rather than reactive, it is extremely important to have an early decision regarding levels of supplies required in a particular theater and the likely duration for which required. Without such established standards to use as terms of 'reference' it is virtually impossible to realistically determine requirements for items of supply, storage, maintenance, ammunition and other utilities, as well as the resultant unloading capabilities and storage facilities needed in a particular theater. Unless these standards are prescribed well in advance, the logistics system has no grounds for the challenging requirements placed upon it. The logistician must also remember that " staying power" in combat is the ability of the forces to remain in the lethal battlefield until the battle is won. The logistics back up must ensure that this staying power is not impaired. To be capable of doing this the close support logistic units will have to be flexible, movable and responsive to the needs of troops. In so far as logistics is concerned, the whole "fighting machine" must be constructed as a unit adapted, in kind and strength, to meet such tasks as may be imposed. All this means is that the logistic support needs to be more adaptive and innovative. (".. .. ... the flexible employment of forces is the central task in directing a war, a task most difficult to perform well. . .. .. flexibility in command can be realised only through the discovery of order, light and certainly amidst such circumstances peculiar to war as confusion, darkness and uncertainty." -Mao Tse Tung). "Flexibility" in logistics support is an absolute essential for a commander to exploit either his success or an unforeseen opportunity during the course of an operation. The Germans during the 1st World War could not sustain their offensive on the Western Front despite new flexible tactics they employed because their logistics did not support these tactical innovations. The logistics plans should be carefully scrutinised to determine which aspects contribute to flexibility and which may detract from it.

To support flexibility in plans there should be flexibility in forces, -availability of troops and organisations as well as in storage and other facilities to meet changing needs. A two-pronged approach helps in giving adequate flexibility. The logistics complement is split into two parts. Namely, "Strategic Administration" and "Tactical Administration". Strategic administration should cover the entire system of supply and evacuation between the main production centers, base installations and the rear echelons which are integral parts of the field armies. It is a system, which obviously increases greatly in complexities with longer lines of communication between the main and forward bases. It is an unavoidable situation because it is essential to preposition adequate stocks at the least vulnerable strategic points. Such pre-stocking, however, must be dispersed and must be as complete as possible in order that the disruption of any one base or group of bases does not prejudice the war effort. 'Local tactical reserves' must be held in the theatre against any temporary breakdown in the supply-machine. Such reserves must be held dispersed as far back as possible at the level of the highest formation in the theater. In short the distribution system be such that the stores should move forward to progressively smaller and smaller installations until finally they are delivered to those who require them. Ammunition logistics support is likely to be the biggest problem, because the sheer size, volume and weight of the items required in addition to the types required will generally be difficult to foresee in totality. Use of 'Containers', as discussed earlier, can help in this situation because the containers can be diverted to a place where a particular type of ammunition is required, with minimum effort.

Another problem area is the provision of repair and maintenance backing. Very often repair and spares facilities are created in such a manner that induction of different equipment causes breakdown in maintenance services. This causes delays and the troops are denied usage of their integral resources. Generally the reason for this kind of situation is the fact that the Services are taken into confidence rather late. In this era of high technology expertise with regard to a particular equipment takes time to build up and so does the placement of backup spares and repair facilities. In order to have "flexibility" with regard to the type of equipment it is essential that indications of intentions is given much in advance to the planners who are responsible for providing logistics cover to the troops. This will ensure that the logistics infrastructure is able to accommodate new equipment and new type of logistics problems that are likely to arise with the rapidly changing technology that we see about us. We cannot afford the luxury of designing a logistics structure for each new technology that comes along. The system must be flexible enough so that it can adapt to new situations without distracting its capability of meeting the needs of troops in the field. It can happen that an operational commander may lack confidence in his technical support initially when a new equipment is introduced in his theater and the artificers haven't as yet gained enough expertise in providing suitable repair cover. He may not like to fully exploit the equipment in such circumstances. It will be better if initially a highly mobile " core group" of technicians is created in a central location to be ready to move at a short notice and be trained and equipped well enough to deal with a given situation. Such 'core groups' can be made "heli-borne" so that their reaction time is considerably less. Operational commanders on seeing such support for newly introduced and complicated equipment will be able to exploit resources at their command more enthusiastically.

Dispersion

Traditionally, logistics units are so sited that they are close to each other and are placed in a manner that is convenient for the combat troops to draw their requirements from. Dispersion of these units over a much larger spread has generally been given lower priority because a vastly spread out administrative area is that much more difficult to protect. In today's world where technology has brought in weapon systems of tremendous accuracy and with a much more lethal effect it has become essential that these units be well dispersed so that a single air-strike or a guided missile does not lead to incapacitation of a particular support service. Where essential stores, ammunition, fuel and those items that are identified as vital are concerned it would be appropriate to have 'multi-point' storage even within a unit. That would provide dispersion and even make issues faster when more than one unit comes for collection to the service installation.

In the case of those units and formations employed on IS duties, low intensity conflicts and in counter insurgency roles dispersion of logistics units and stores will be more important. For various reasons it may not always be possible to make use of the local resources as the population and vendors of the area may not want to antagonise the insurgents. In such situations having lighter combat support units that are self contained in terms of protection, transport and communications can play an important role because lack of a set pattern will make it that much more difficult for the hostiles to cause damage as these units could quickly pack/unpack , perform their tasks and move on to the next location. Security of the communication channels will be essential because dispersion will make inter and intra unit communication prone to hostile action. Buried cable in addition to radio will be required to ensure that there is no break in communication. Where considered to be operationally necessary communication channels can be duplicated. Locations of more important stores could also be changed frequently. This will deny hostiles or saboteurs chance of causing damage to support service. Creation of a "combat support network" where all essential elements of material support are inter-linked by means of computer based system can be a big help to operational commanders. For predetermined and already identified repair and spare support helicopters and even air effort can be allocated before troops are launched. Besides ensuring safety and dispersion of such costly stores such an arrangement will also provide a much higher degree of flexibility.

Logistics Momentum

"Logistics momentum", as such, is somewhat different from the momentum of the combat forces that can gain and sustain their operational speed for a limited range with their "first-line" resources. Logistics momentum becomes of importance in the roots of our logistical system. It spreads all through the broad fields of procurement and distribution, and it reaches its most critical point in the operation of large forces in combat. In a major operation, the logistical preparations develop an actual physical momentum of the "means" of combat; a momentum which must be recognised in the planning and conduct of such operation. It takes time and great effort to build up the advanced stockpiles necessary to support sustained operations. It is all the more difficult in a country like India where troops may simultaneously have to operate in vastly different terrains and areas where road and rail networks may or may not be adequate or different rail gauges may be involved. There could also be wide gaps in the 'authorisations' and 'holdings' of combat units needing to be closed as a first step towards operational readiness before we even get to think of providing any momentum to the build up. Another danger to watch for is that our supply system may be geared too closely to peacetime operations; and that it may not either technically or organisationally be prepared for the very great changes that a war-like situation is likely to bring.

In peacetime, movements are of less volume and are more predictable than they are in wartime. The major movements and transportation can thus operate effectively with a high degree of centralisation. However, in wartime with greater volume, greater urgency and less predictability, the routine inter-theater systems cannot handle all the necessary work. An important aspect of this situation is that in a short intense war there is no margin for an error of judgement, and the penalty for delay in material for continuation towards the aim is seriously linked with effective and efficient logistics backing providing momentum to the combat forces. It is seldom understood that during the course of a logistical buildup the material tends to move at such a fast pace and on such diverse transportation modes that it is virtually impossible to reverse the flow. It is very difficult to change its rate or direction unless we set up an efficient, accurate and rapid system of logistics controls and communication. In the absence of such controls the momentum of the personnel and material in the pipeline will be such that it continues to flow regardless of need or of high command orders. The excess spills out the end of the pipe in the combat zone and creates confusion, trouble and waste. Another problem area is that this excess becomes a burden to the combat forces by reducing their mobility and flexibility because the stores cannot be left behind unguarded and unattended. The unwanted materials clog access to the wanted material all through the system of transportation and the depots. The organisation of the transportation and movement systems is closely related to the 'mobility' and 'flexibility' of the combat forces. Establishment of an intra-theater system under a 'resource manager' more directly responsible to the theater commander can ease the situation. This resource manager should be in full picture of an operational situation and the higher commander's plan so that he has sufficient reaction time to alter the flow of material when required. Things will become more manageable once we have 'asset visibility' and computer based controls.

Feasibility

> Many Generals have failed in war because they neglected to ensure that what they wanted to achieve operationally was commensurate with their administrative resources; and some have failed because they over-insured in this respect. The lesson is, there must always be a nice balance between the two requirements.
>
> Field Marshal Montgomery
>
> (In " The Memoirs of Field Marshal the Viscount Montgomery of Alamein")

Not only are strategic and tactical plans limited by the feasibility of logistics support; logistics plans themselves are subject to the capabilities of the national economy, the availability of other resources, and the limitations of secondary requirements. There will never be enough money to meet all the requirements of the armed forces. In "peace time" when the Strategic and Operational plans are designed and developed highlighting the material needs of the forces to support those plans there will always be certain things which the government of the day may think to be more important than the materials required by the Services. They must at this stage define their priorities in so far as logistics are concerned because if a particular strategic plan is developed with 'unsure' supply back up it could create a wide array of problems in due course of time. All options including exploration of civil trade and industry to meet specific military needs must be examined. In sum total the feasibility is to be ensured at some point of time while concretising operational plans. That, the national economy and the military strength are inter-linked has been discussed elsewhere at length. At this stage it will suffice to say that the logistics resources are almost always limited and must be concentrated in the best way to achieve the primary mission. Therefore, one great thing to which we, the logisticians, should at all times apply our thoughts and brains is the expansion of the power of material and personnel without increasing either so that an operational commander's plan becomes either feasible or it could be modified within the framework of resources available. Never the less, in all cases the costs must be considered and the least expensive means consistent with the primary purpose should be chosen. Otherwise, in due course of time it might become difficult to sustain the forces. It is advisable that the ratio of secondary to primary requirements is kept low. The production capability of own industry, both in the 'Defence sector, and the 'Civilian Industry', or our capacity to obtain the same in the case of an imported item may be as important as battlefield performance of the combat troops. Any operational plan without a due consideration of all the aspects of procurement, placement, prioratisation, and plans for induction in the theater of operations can cause tremendous problems at all levels.

Feasibility is not related to strategic or operational plans alone. Capability of the logistics infrastructure to support these plans must be examined alongside. Even at the level of a divisional operation it will be pertinent to remember that 70 to 80 ten thousand liter tanker-loads of fuel will be required to replenish the fuel burned by an armoured division in a 50 kilometer attack/advance. Other replenishments like ammunition, spares and rations will require over 200 vehicles and tankers. All operational planners should consider these aspects before they come to the conclusion with regard to feasibility or otherwise of their plans.

Timeliness

" Time is the irretrievable factor in war."

Napoleon Bonaparte

The operative requirement here is that the logistics support must be made available in the desired quantity at the right place and time to allow the operational commander to carry out his task. The system must be flexible, supple and adaptable to support quick changes in the operational situation regardless of any enemy interference. Time is paramount. This important principle can make all the difference between success and failure of a tactical plan. The time aspect is one of the most critical elements, since resupply of units must occur at the decisive point. " To get there first with the most " as General Nathan Forrest of USA once stated is no longer sufficient. In today's environment there must be qualitative factor included in this time equation to ensure we can adequately rearm and refuel the forces. (It is said that the British Army was able to escape from Dunkirk because the advancing German Panzers were running short of fuel). Time will make the difference between success and failure on a fluid, high-operational tempo battlefield. This is especially true with today's weapon systems, where the complexity of technology, heavier caliber of the weapons, and consequently heavier ammunition expenditures have significantly increased the logistics requirements. Adequate and timely supplies provide the commander an opportunity to gain the tactical advantage, where too little would severely restrict his options. The timing aspect must focus on attainment of the objective. Timing is critical to maintain a forward impetus and continuity of operation. Field Marshal Rommel's failure in North African campaign during 2nd World War can be attributed (in part) to the lack of responsive and timely supply of his material requirements to maintain and sustain the tempo of his operations.

In order to provide timely material support options are often limited because if we start positioning stores in anticipation of a particular development it might give the whole thing away. One problem with logistics is that the sheer volume of the stores, size of the machinery and plants, requirement of space and the means of transportation etc cannot really permit secrecy. A buildup of these elements will be indicative of the operational plans. Nor, can a commander afford to delay such buildups if he wants his operational plan to succeed. There will always be a need to have a balanced view and a compromise between the two. The length to which the allies went to hide their intentions about Normandy-landing from Hitler and the effort it involved is a matter of recorded history. It may neither be feasible; nor, economically viable to go to such lengths. It would be enough to understand that the logistics tasks should be accomplished with celerity that retains initiative and takes full advantage of opportunities of the moment. Timing must be related to the objective, whether in high level procurements or a tactical supply. The timing of an action is likely to be as important to results as the action itself. This means that the logisticians must be capable of reading the operational situation as it develops and be ready with their support. They should be capable of reaching meaningful conclusions from various intelligence inputs. Commanders need to appreciate that this information is equally important to the logisticians, as it is to the operational staff. But then the logisticians hardly ever are taken into confidence. It is an area of neglect that requires immediate attention. Movement of troops and materials must be scheduled and conducted with a deliberate speed that avoids hastiness that contributes to confusion and congestion and the tardiness that weakens support and delays action It is essential to work out this aspect at a much greater length during peacetime exercises.

Timeliness is not related to equipping the 'field forces' alone. Production, procurement, conveying the resources to the desired destination assume equal significance. Equipment to support our forces is produced after the need or requirement becomes apparent. Naturally there will be shortages as a lag between the establishment of requirements and availability of sufficient assets exists. We need to streamline our procedures and methods of procurement, including time expended in obtaining the government sanctions. No matter how so ever hard we try and whatever money the nation spends the road, rail and air network in the country can never be adequate for an ideal operational deployment of men and materials. All that we need to understand is that logistics is relative to time, place and circumstances. Depending on the criticality of the situation, relative importance of a particular system for furthering the aim all effort must be made to position resources at a commander's disposal. Preparedness can never be absolute, only relative to the time and place of possible conflict. Therefore, logistics activities in one area or command must be related to the others to meet requirements in a flexible manner so that maximum benefit is drawn from available resources.

Sustainability.

In classical terms sustainability is described as," The ability of forces to maintain the necessary level and duration of combat activity to achieve their objectives. This means having sufficient personnel, equipment and stocks on hand and having the ability to resupply and reinforce on a continuous basis".[5] It is normally expressed in terms of days. It reflects a commander's subjective assessment of the overall capabilities of his command to sustain military operations; and the ability of forces to maintain the necessary level and duration of combat activity to achieve their planned objectives. It represents the balanced capability for all logistics and combat support functions, which provide the force staying power over time. It includes the force structure, pre-positioned and war reserve materials, prescribed loads and operating stocks, wholesale sustaining and the country's industrial base which in their totality comprise Army capability to attain the desired mission. it means a very large number of support vehicles, technicians, storekeepers, and others who are to man these units, and be ready all the time to sustain the forces committed in any operation. The whole thing makes the issue of sustainment ridden with too many ifs and buts. Because larger the logistics load lesser is the mobility and flexibility afforded to a commander. To move so many support vehicles, in addition to combat assets of the division operating along the attack axis, requires several passable routes/roads and staging areas to down load the cargo. And where required to repack the stores for further transportation by means that are available in a particular terrain. These are lucrative and highly vulnerable targets, particularly to the 'enemy air' and while replenishment is going on during a lull in the battle. The issue becomes even more critical in respect of mobile forces because the unit basic load's replenishment must occur prior to the deep attack force being engaged by the follow on enemy forces. The British Army has calculated that an armoured division will require nearly 4000 tons of ammunition of all types per day. By far the largest proportion will be artillery ammunition. Guns will be expected to move several times a day to have a chance of surviving counter artillery fire. If they leave ammunition behind, it must, if possible, be picked up. This compounds the already huge problem that the transport will have in servicing the current gun positions and stocking the next, as well as clearing those just vacated.[6] This is one example highlighting the magnitude of the task involved.

System of Sustenance

The biggest nightmare of the logistician is to sustain the combat troops both during the period of relative calm and actual hostilities. Whereas the quiet in the area could be deceptive and misleading thus giving no clue to the enemy's intentions own commander would like to be ready for any eventuality. This means that in so far as the logistician is concerned he has to cater for and plan to sustain the supported forces for an in-determinant duration and in an, largely, un-specified manner. In order to sustain troops when there is only a 'waiting and holding' environment not knowing what the other side is contemplating it becomes necessary to locate some stocks at intermediary depots to ensure timely issues. The "war wastage reserves"(WWR) have also got to be positioned in a manner which ensures continued support to the troops. Technically, WWR is defined as a reserve intended to cover the wastage of items of equipment, ammunition, vehicles, rations, and such like stores during operations up to a maximum of first six months of a campaign or until the regular supply catches up. These reserves are required to be dispersed both from the operational and administration point of view. A portion of the WWR is placed at the disposal of Head Quarters Command who locate them at convenient locations. Sustainability of formations in a particular theater and their likely operational role dictates the stocking policy. This stocking policy is based on following factors:

- Adequate dispersion: Necessary for two reasons; one, makes enemy's task of dislocation and disturbance that much more difficult. Second, gives flexibility to the troops in the form of that many options for drawing their requirements.

- Availability in close proximity of the field formations. Taking our strategical and tactical resources as close as possible to the fighting troops considerably reduces their turn around time and consequently reduces the need for them to carry extra load.
- Holding capacity and capability of various echelons. Larger the number of dumps greater is the need to provide them security. The stocks should be within the handling capability of the installation. Very often stocks have to be kept at convenient locations. This is generally so in the case of unwieldy, heavy, and those sensitive items that need frequent checks and inspections.
- Financial constraints: Stocks held as reserves mean locked up capital. If these are to be dispersed over different locations, then we need more quantity. Financial capacity to maintain such stocks will govern the decision with regards to the level of stocks.
- Wastage pattern in the past: Wastage pattern for different items in different terrains will be a major guideline. For example , spares and accessories will create a greater load in deserts. Or the quality of road network and means of communication in a particular region will define a pattern of wastage of certain items.
- Availability of local resources: Plays a prominent role in deciding the stocking levels for sustainment. Every effort should be made to identify local agencies who can meet normal requirements. Besides saving on creation of departmental capacity such an exercise helps during operations. Only risk is that an impending operation can create artificial scarcity.
- Communication System: Where means of communications are primitive or prone to breakdown stocks are held in a much larger measure. For example, the 'road closure' in the Leh region necessitates winter stocking during fair weather. Likely

duration of such breakdowns will be a guideline for the stocking policy.

- General environment in the region and changing trend in the government's (foreign)policy: Things like "forward posture" define the stocking policy for a particular region. Building up reserves of high caliber artillery ammunition, and tinned rations etc could cause a lot of problem in due course when these things do not get consumed during its shelf-life. Bringing them down to the plains for speedier consumption is prohibitively expensive. Therefore, such items should only be stocked in the area of deployment up to the prescribed scale of consumption for a specified period. The remaining stocks be preferably held at the 'rail head'.
- Evaluation of previous war experiences: Generally provides guidelines for the 'war wastage reserves'. However, with the advancement in technology and improvement of communications network such experiences tend to get outdated, more so when the nation hasn't been to a war for a long time. It will be prudent to learn from the experiences of others as well so that we could consider an increase or decrease of various scales.

Methods of Sustainment

In certain armies, including our own, usually 'Push System' is advocated when operational scales and wastage rates are well established. The purpose of this method is to provide a form of automatic re-supply to meet the anticipated consumption and to support the prescribed scales. With experience gained over the last so many years we could design specific packages for various terrains, types of equipment and nature of operations. Push Packages should be based on the latest experience factors available and should be scrubbed to exclude those items that are "nice to have". The quantities and contents should never be more than the " provision data". If we bring in containerisation then it will be easier to design the contents. These packages should only be used as interim relief measures. The Americans used the system extensively in Viet-Nam. Their experience and the problems encountered by them and as recorded in "Viet-Nam Studies" by Lt. Gen. Joseph M. Heiser, Jr. are appended here in:

1. **The inability of "inexperienced unit supply personnel" to sort, identify, and properly issue the volumes of items received.** In many cases, the original "push packages" and their individual contents are identifiable only by shipment number and stock number (part number). Therefore, highly skilled persons are required for placing these items in the proper bins to make them readily available for use.
2. **A lack of secure warehouse and storage space and sufficient logistics units at depots.** The size of individual push package increments arriving initially was too large for the discharge capacity, storage facilities available, and the capabilities of the supply units at the unloading areas.
3. **Fast changing destinations.** In some cases, unit destinations were changed without changing the destination of the package, causing delay in the receipt of the package by the intended unit.
4. **Delay in reverting to the normal system.** Inexperienced personnel became accustomed to the "push" philosophy and failed to evaluate measures for transition to the normal "pull" system.
5. **Inadequate Service Personnel to handle stores.** An enormous amount of material shipped to our forces on operations was not readily available to them because of poor visibility of assets in the theater, one of the lingering effects of the misguided "tooth to tail ratio" debate.

In order to safeguard against such problems it is recommended that these packages be developed on an austere basis and include plans which permit transition to normal supply support as soon as the situation permits. Push packages should make maximum use of standard containers with stocks pre-binned and accompanied by locator cards. Push packages be approved only as an interim relief measure that permits the pull system to again respond to the unit's needs when the situation is more or less stabilised. It can happen that the actual waste is much less than the approved size of the package. Units normally are not likely to high light their sudden flow of plenty. It is for the service heads at various levels to ensure that no surpluses are created. They could adjust such extras within their own formations and inform base to suitably modify next consignments.

Another factor that can help is having "Closed Loop Support". In this a piece of equipment is, as far as possible, retained within a prescribed theater of operations. Necessary repairs are carried out there only and the equipment re-circulated. The objective will be to ensure response to the needs of operational units to exert more effective control of critical serviceable and unserviceable assets in the logistic pipe-lines to reduce the backlog of unserviceables at all levels. This means that the repair support from dedicated workshops will have to be as forward as possible in order to ensure that there is no or limited backward move of the repairable equipment. With the modular concept catching up it should not be very difficult to undertake instant repairs. Cannibalisation (it is the use of parts from one disabled item to repair and restore another) is another approach that can be of very big help in sustaining forces in mobile operations. Only the difficult cases be back-loaded. This would ensure timely availability of repairable assets at depots and maintenance including overhaul facilities. A commander will also be able to establish asset- control of the flow of controlled and short supply items. Basically, the term sustainability carries a connotation of supplying of material needs. One factor we must take into account is the fact that different operations, terrains, operational considerations ie, is it a holding operation or offensive will need different kind of consideration if it has to be sustained efficiently and competently. These aspects of logistic considerations as they relate to certain operations are broadly discussed herein:

1. Desert Operations.

In defensive operations and also during 'peace time' the formations are deployed around nodal points. They will need "tailor made" logistics units meeting specific requirements of the formation. In offensive operations armoured/mechanised/ RAPIDS will require logistics support that enhances their operational capability and manouverability. This means that logistics units will have to be compatible in mobility to the supported formations with light and essential stocks. In the deserts the frontages are vast, dispersion is the norm. Space for manoeuvre is extensive, and the requirements for concentration of the force is dictated by speed in the shortest possible time. The troops, in all likely-hood will function as 'combat groups', and 'combat teams' away from the main force with the ability to move and concentrate at a point of decision in short time to achieve their objective. In such a situation the logistic support is required to be flexible enough to cater for these quick concentrations and dispersions to enable commanders to carry out their plans effectively. Handling of heavier ammunition and assemblies for the tanks and the guns is a labour intensive task. Scarcity of the civilian population in the desert makes availability of labour-force a problem. Due to lack of vegetation, camouflage and concealment and also dispersion of the logistic units is difficult. The spread of the desert battlefield puts units further apart, makes them more difficult to find and requires the resupply vehicles to travel further. Consequently, in a 'fluid battle situation' move of the transport columns is restricted to the hours of darkness. Communications are also subject to disruption due to heavy dust storms and enemy action. (During operation 'Desert Shield' the Americans found that Iraq had made tremendous use of line and cable that was buried in sand. Even if the American Air damaged the cable at certain places the Iraqi troops managed to repair it during the night. The obvious lesson seems to be that it might be more prudent to rely on good old line communications while preparing defensive positions in the desert.)

Unless "air superiority" is obtained, the lack of natural cover makes movement in daylight perilous. If the logistic system has 'air lift' capability and a matching mobility with the mechanised forces the task becomes relatively more reliable. Navigation in the desert is not easy. During operation 'Desert-Storm' the Americans used hand-held or vehicle-mounted navigation aids often linked directly to a satellite. This helped the logistic vehicles in finding the supported units across the Kuwait desert.

The worst is that the desert reserves its harshest treatment for equipment. All moving parts are subject to the abrasive effect of sand and rocks.(Sand ingestion was a major problem for many heavy armoured weapons and for helicopters operating out of unimproved fields. Many sophisticated electronic and mechanical systems either required modifications of some kind, or required major changes to maintenance procedures and far more rapid replacements of certain key parts and subassemblies. Fuel requirements were often higher than anticipated, and the logistics load was often much higher than was anticipated before Desert Shield[7]. Even helicopters under performed unless adequate 'sand' filters were provided- an item that was not catered for in abundance) One can really never be enough prepared for sustaining troops in deserts. For efficient sustainment we could create "Quick Reaction Logistics" teams. These teams should have mobility commensurate with the mobile operating forces and carry sufficient spares, oils and accessories and provide replacements on "clean exchange" basis. Such teams should be formed even in peace time (particularly during exercises) for performing these tasks. This will help all concerned in arriving at a realistic list of tasks to be performed, and spares etc to be carried by these teams during operations.

Experience tells us that as much as 40 percent of the vehicle tyres of a unit can be flat after a long cross-country move. Every vehicle has to carry an extra fan-belt. The consumption of the spare parts will be almost double. To make matters worse every thing that can expand in the searing heat of the desert day, or contract in the cold of the desert night, will do so. This means that Engineering facilities for forward repairs and fault diagnosis are required to move in concert with the fighting troops. Down time of the equipment will largely depend on the assurance levels of the back up inventory. Therefore, items like engines and gearboxes showing high wastage be stored as far forward as possible. But it could also mean an increase in the "weight" of a unit, which ideally should be as light as possible. Cannibalisation, as discussed earlier, has merit in so far as supporting maneuver-oriented units are concerned. A system where hulks and disabled equipments are reported to higher headquarters must become every unit's charter. A commander can send a "scrounge mission" as and when the situation permits him to do so. All the disabled and discarded equipment be stripped of its serviceable parts and these be used to resurrect those that can be put back on road. A selected range of spares be so placed that it is within an easy turn-round distance of the supporting material organisation. Consignor to consignee service can also be adopted to cut on the lead-time in making spares available in the shortest possible time besides ensuring a higher level of reliability. This also means that the latest communication facilities must be provided to the service units as well. The envisaged logistic support system should have the capability of sustaining fighting elements by enhancing their staying power on the battlefield. Such a system should provide the following facilities:

- Quick response; this means that the logistics system must be geared to support the vast distances which formations will be required to cover, forward and laterally so that the flexibility of the higher commander remains unhindered; In this forward stocking in smaller set ups can help in sustaining the troops.

- Be flexible enough to support from any maintenance base; even though the repair support from the dedicated workshops is likely to be available it would be prudent to cater for such assistance from anywhere in the theater.
- Experience has shown that long range ammunition consumption almost doubles up while close quarter ammunition rates reduce by a third. Realistic ammunition targets be worked out and positioned accordingly. Insides of modern missiles can be degraded by high temperature or the direct rays of the sun. These should normally not be removed from their air-conditioned storage space unless likely to be expended quickly.
- Water sources in the desert may be non-existent or polluted. In built carriages for water and the capabilities to locate sources "with suitable drilling rigs fitted on tracked/semi-tracked vehicles for Engineers" for locating and developing water sources assumes special importance. Ideally a special group of engineers be organised duly equipped for establishing water points.

2. Counter Insurgency Operations

The conflict is quite different from that for which troops have been trained. For that matter, it is at variance with combat dynamics upon which the Army's logistic doctrine is based. The operations are conducted essentially by small units, like battalions, companies, platoons and similar forces in constant pursuit of an elusive hostile; unlike conventional war there is no fixed terrain objective. There is no neat, linear division between enemy and friendly forces, no rear boundaries. Consequently, there is no specific Army Area, nor a Communication Zone, per se. This results in a situation where at no time do we really have "secure" depots, storage facilities, service areas or supply routes. The relative degree of security will, therefore, vary from time to time, and place to place. Attacks by the 'hostiles' or efforts to penetrate into the logistics facilities and operations at all levels tend to pose a problem. There could be major ambushes of supply convoys; harassment by small arms fire; and even rocket & mortar bombardments. Since, these operations are usually over inhospitable terrain and the operational commander has to face an all round combat threat he cannot really site his logistics units where they could be most effective, unless he has resources to ensure their safety at all times. No capable and realistic army can promise this. Physical security in CI operations and hostile areas thus becomes a major consideration than what it is in a war characterised by front lines and relatively secure rear areas. Convoys will need protection from hostile elements. It would be more prudent to keep the "stock levels" small and austere so that they present less lucrative targets. Best policy will be to tailor resources to the combat posture in consonance with the theater; prune both the unwanted and non-moving inventories. Both the service units and the combat troops have a tendency to hang on to too many things fearing the unknown. The use of helicopters for logistics supply has proved to be successful in Viet-Nam and more recently in the Falklands and Kuwait. But they can move only a fraction of the quantity of stores, fuel, and ammunition required in a major battle or to sustain a major force. This situation is most unlikely to alter in the near future. Therefore, at best, the helicopters can only be a useful back-up to normal methods of supply and be used only for moving urgently required spare parts, for supply of ammunition to isolated units and, in particular, for rapid evacuation of casualties. Traditional methods of transport will have to be used to move the main bulk of logistics supplies. The functionaries in the logistics units must be alert and alive to the operational situation and be proactive rather than reactive. For example, on learning of an engagement they should start planning, preparing and packaging replacements and replenishments on their own rather than wait for a demand from the user. Since they will normally be located close to a headquarter flow of information can be coordinated where some specific task or activity is envisaged.

At unit level there will be three areas of concern- preventive medicines, medical treatment and casuality evacuation. In certain parts of our country the troops could become ineffective due to various local diseases like malaria, dysentery and such like maladies. The sustainment package should contain the prescribed preventive medicines. If the situation develops into an armed conflict there will be casualties. These be treated as far forward as possible so as to enhance their chances of survival. Modern medical science and more effective antibiotics, portable surgical and resuscitation kits have made it possible for the doctors in the Regimental Aid Post and the forward field ambulances to much improve the chances of a man surviving what in earlier operations would have been a fatal wound. But there will never be enough medical personnel available; therefore, necessity for training of more personnel in 'first aid duties' cannot be over emphasised so that a casuality can be sustained and kept alive till evacuation to an appropriate hospital can be arranged. A rifleman per section should preferably be trained in providing first aid.

A major bottleneck is that in our system the units have to depend on the formation's integral service units. At times it happens that a fragmented formation gets inducted (particularly in counter insurgency operations) without adequate integral support. This necessitates their being dependent on the conveniently placed logistics units. The trouble is that this kind of service involves a long drawn and perfectly avoidable procedure. Service units cannot undertake the responsibility on their own, unless the 'Service Heads' are willing to give a go by to the rules. But even then their hands are tied, even if the logistic units are willing to be helpful they cannot really and effectively do so because the ' audit authorities ' may not allow them to do so. The Service units are auditable. (The system is that when a unit moves out of area of supply of one Service unit they are required to cancel their demands, obtain a certificate to this effect and place fresh demands on arrival in their new locations). In CI operations- of the kind that we have been having of late- there is often a brisk move from one formation to another. This causes serious problems to the troops insofar as their logistics support is concerned because the newly inducted unit, even if it places its demand in the new set up it will be last in the queue. A pragmatic approach by the 'Support Service Commander' can ease the situation, and it often does. We definitely need to change our existing procedures and policies on the subject (horizontal dependency and supply of material support in such like situations has been discussed earlier and elsewhere). Hopefully, the inventory automation activity will resolve this problem to a great extent.

3. Jungle Operations

A jungle operation creates special challenges to the logistician. The topography will normally create difficulties that hinder activities like construction of storage-sheds, hard-standings for vehicle and other equipment repair workshops, transportation and communication. The environment, at the same time will sometimes facilitate the "opposition's operations". Many of the techniques and assumptions that are accepted as valid in conventional warfare do not apply in the harsh, primitive jungle environment, and the isolated support enclaves as the Americans learnt (the hard way) in Vietnam. The first of these problems is normally the limited infrastructure in the theatre of operations. The terrain is sparsely populated and roads rarely surfaced, limited and vulnerable to closure by mining by the hostiles and damage due to the weather. One might have light aircraft landing grounds and temporary 'air-strips'. But these will rarely be all weather and unlikely to be able to take large military air crafts. This imposes an almost 'stop-go' tempo to an operation in the jungle. Since the hostiles are more likely to have local sympathy and support, chances of a complement of troops getting isolated cannot be ruled out. In fact, as the Americans experienced in Viet nam, even the whole force could get isolated. In such circumstances the commander has to impose and enforce some form of rationing. Again, like most things in logistics it is simple in concept but extremely difficult in execution. Where and what all can a commander ration and for what duration. Such unpredictable situations do not in any way reduce pressure on the support services where they are expected to perform. The main problem will be casuality evacuation, their treatment, fuel and ammunition supply. Logisticians need to ensure that at no time the support is a constraint on a major tactical operation. Logisticians must accept that whereas the situation imposes a severe strain on logistic operations it also offers an enormous challenge. History is full of occasions when logistic risks were taken to break-out of this cycle and they did not come off leading to catastrophe. Giap in Vietnam managed to overcome the French in Dien bin Phu and later much more technologically advanced American opposition during a long drawn war. Primarily ,he was more successful because logistical needs of his forces were limited and the Vietnamese troops were capable of making use of local resources to the optimum. In our own environment the troops have been dealing with such unpredictable situations both in the North East and more inhospitable areas of the valley.

An important thing to guard against is that tropical conditions do play havoc with complicated electrical and electronic equipment and significant repair effort has to be allocated to keep the myriad of electronic devices required by a modern army fit for action. During peace-time the emphasis in most of the cases is placed on the combat and combat support arms. Combat service support units and trained technical personnel and the technicians do not get to practically experience the problems which will arise in actual situations. As a consequence, when contingency operations are undertaken serious deficiencies in logistic units and trained personnel may be expected. There is an urgent need to enhance the readiness of the Services as well, so that they could respond promptly and efficiently to limited and often unpredicted operations that come up from time to time. Again, of all the logistic functions it is the Medical Services for whom the jungle presents the greater challenge. Building heli-pads quickly, moving life saving drugs from one location to another, and practicing casuality evacuation etc are some of the, things that require practice and refinement during training exercises.

4. Snow Operations

Snow hampers all movement. It can cloak static logistic sites and make camouflage of ammunition stacks and piles of supplies easier. But it also clogs the routes to and around these sites. Working and fighting in the snow requires a high calorific intake. This means much higher scales of 'troop's rations' have to be provisioned for. Frost bite and Snow Blindness can keep the Medicos busy unless commanders insist on proper clothing and care by the troops. Traditional logistic unit-siting distances have to be adjusted. In this much will depend on enemy's mortars. But, if shielded by high ground, the stockpiles could probably move reasonably close to the front. The problem is to find sufficient flat ground on which to site logistic units where from dependent troops could draw replenishments conveniently. "Vehicle Waiting Areas" and hard surfaces for smooth flow of traffic will need considerable 'Engineer' effort. It must also be remembered that fog restricts visibility considerably, even during the day. Peculiar and special 'snow warfare' equipment requirements have to be worked out much in advance and provisioned for, as generally these are of foreign origin and are scarce. Another important thing is that a particular operational situation may not allow for adequate acclimatisation where troops are concerned. This may mean extra and avoidable work load on the medical services. Assam Rifles, Ladakh Scouts, and Special Frontier Force etc are good examples of how the local population can be employed for combat duties. However, we have not made any particular effort in working out the requirement of additional labour force that would be essential for handling material during operations. Sturdy and physically fit locals can be identified, verified and if possible, be given some kind of training in handling stores and material handling equipment. This work force can be quickly impressed into service when any emergency arises.

The issue of sustainability will always remain subjective and will be based on assessments that do not fully account for the opposition's capability to delay results. If the staff underestimates the intensity of operations and are mistaken about the form they will take, some commodities will run short, and reserves will be consumed. Resultantly, the rate of build up of reserve stocks, to meet the higher expenditure rate of an intense operational engagement, will be critically affected. Particularly, if there is insufficient margin in the transport system to allow both-daily sustenance and an element of reserve supplies to be lifted forward on regular basis. Inspite of the historical records of French defeat at the hands of Vietnamese the Americans severely underestimated the capabilities of Giap when they went into Vietnam. It took them too long to arrive at a proper balance between combat troops and supporting technicians required to sustain their forces. However, past experiences, availability of simulation models and intelligence inputs can lead to an educated guess which can be fairly accurate provided one has been honest enough to make a realistic bid for resources.

Continuity

> In modern warfare no success is possible unless military units are adequately supplied with fuel, ammunition and food and their weapons and equipment are maintained. Modern battle is characterised by resolute and dynamic actions and by abrupt changes in the situation, which call for greater quantity of

> supplies than was the case during the Second World War. Hence the increasingly important role of logistics continuity aimed at supplying each soldier in good time with everything he needs for fulfilling his combat mission.
>
> Colonel General Golushko, Chief of Logistics Staff, Red Army Armed Forces

An operational commander has to be sure of material support for his task in hand, then only can he plan deployment and sustenance of his troops. There could be any number of reasons that could disturb the continued material support. It could be due to enemy action, or as it happens in high-altitude areas a landslide is more than likely to hinder the normal supplies. In a holding or defensive operation plans for continuity are relatively easier to make. The real problem relates to those operations where either an advance is contemplated, or in CI operations where the hostiles are more than likely to cut off the line of communications. In such situations it is difficult to depend on local resources either. Therefore, a higher commander, in consultation with his technical advisers, needs to lay down scales of various stores to be held in a particular theater for a prescribed duration. In this, factors like availability of air dropping, helicopters, and local resources will also influence the decision. 'Service' units must be organised in such a manner that no fundamental change is necessary in responding to an emergency. Problems come up when these units have to provide detachments far removed from their base on almost permanent basis and commanders expect absolute efficiency. Since these units are hardly ever designed for this type of contingencies; there are serious repercussions of such squandering of already scarce logistics resources. Insufficient manning levels lower their efficiency, and inadequacy of own troops makes them ideal targets for infiltrators and saboteurs. Disturbances so caused affects their functioning and ultimately continuity of their support.

Continuity also relates to continuation of plans and policies, as frequent change in either orders or directions in making material support available to the troops can have an adverse effect both on the morale of the combat troops and that of the Support Services. It is no secret that continuity of experience provides the basis for the continuation or modification of an old policy or for the introduction of a new one. Presently, having cut the tail beyond reasonable proportions we do succumb to makeshift arrangements where our service units are concerned. Since these unorthodox and retrograde actions of operational commanders haven't been tested in an actual war it will be unfair to make a prediction. One could say with certain amount of certainty that most of these adhoc arrangements will fail when a real enemy puts them to test.

Another area that needs a constant feel of the pulse in order to ensure continuity is our production and procurement systems. One finds that we are too hamstrung by policies and procedures to ensure continuity. If fifteen quotations have to be called for a particular item then one has to approach that many suppliers, even if there are not more than two in the vicinity. Insofar as the system is concerned lack of infrastructure in a particular area of troop's deployment is no reason for not going by the 'rule book'. If the user suffers , so be it. Audit will take a serious view of this violation. This kind of rule bound approach has created a whole network of dummy suppliers who only provide quotations so that the procedural requirement is met. Since government's money is involved risk taking is not considered to be advisable. Consequently, over the last few years complacency seems to have set in. Delays, failure to meet prescribed standards, unilaterally changing specifications by the suppliers are common place. Allowing this approach to continue is not a healthy portent if we have to ensure continuity in our material support. This aspect has been discussed while considering different facets of logistics. At this stage it would be prudent to say that the perfection of logistics organisation and the development of production models of essential equipment should bc a continuous process in peacetime to prepare for war.

Accountability

Every logistics task must be the clear responsibility of someone, and each person must be responsible to someone for performance. It is required because of the great increase in the value (essentially reflected in cost), but also in operational readiness of the equipment, be it armament, communication, ammunition, missiles etc being handled by the front-line troops. We must have a system that requires a minimum inventory and redundancy, and still meet given standards of operational readiness. Accountability should invariably encompass quality control as well at every stage of production, in the development of supplies and equipment and in training of personnel. All concerned should know that excellence in performance of duty is paramount as the state equips its troops at considerable cost and effort. While improvement both in the quality of the product in the hands of the troops and technical support be a constant quest, change for the sake of change, with no significant improvement in quality will be a serious drain on the logistics system. This could also spread the support element over a larger number of service units, thus diluting accountability. In our system with different heads looking after maintenance, spares, rations, and other things necessary for an operationally committed force, there is more than a fair chance that when something goes wrong it will be difficult to identify as to who actually is accountable. Since the manufacturing units are not under Army control there is considerable lack of accountability. It has happened that a box containing fuzes for artillery ammunition exploded while stacking in a unit. During investigation it was found that somebody had been careless in the ammunition factory while screening 'X-ray films' of the safety mechanisms of those fuzes. He had failed to observe that a "safety ball" was not in position at the manufacturing stage. This ball is normally expected to release the striker when the 'fuzed' projectile is in flight to the target. In the instant case the striker had no such restraining medium, and had hit the initiating explosive due to movement of the box while being stored. The person responsible for this negligent act was also identified. But nothing happened because the higher management of the concerned ammunition factory was more worried about its reputation instead of the safety of their customers. Resultantly, even remedial measures to prevent recurrence are also not initiated.

Maintainability

Maintainability has different connotations. First, it entails maintenance of troops deployed in different types of terrains, operations and duration for which maintenance is required. In inhospitable areas where road communications are either primitive or susceptible to vagaries of weather, landslides, snowstorms, or enemy interference troops may have to be maintained by air for long spells of time. Second, maintenance of equipment. We will discuss the maintainability factor in so far as it relates to the troops before examining it in relation to equipment maintenance.

Troops have to be maintained on a daily basis. Daily maintenance means ammunition, food, fuel and other supplies needed to maintain operations on daily basis. This will be based on the size of the force(dependency), the anticipated intensity and type of the operations. Ammunition resupply will depend on its expenditure pattern, fuel on the consumption pattern, vehicles and equipment on the number needing repair or replacement. Almost the only commodity that can be provided with guaranteed accuracy will be rations. Supply of medicines will mean repackaging if the same are to be dropped by air. Maintenance thus implies a real test of the services.

Maintenance by Air

It is expensive in tonnage, as approximately one third of the weight is supply dropping equipment. Breakage and losses are also high. To collect large drops takes a long time. Combat troops will be necessary to protect the personnel collecting the drops and also, possibly as labour. Time spent circling over the dropping zone, and the fatigue of pilots can lead to errors of judgement and stores landing in enemy hands. In actual battle situations air superiority will be a prerequisite before air maintenance is contemplated. Helicopters can play a decisive role in ensuring a high level of maintenance by air.

Maintainability of Equipment

The Services have a vast diversity in equipment holdings, both in range and depth. Larger systems like ships, air crafts, tanks costing crores of rupees are at one end of the spectrum and minor equipment like telephones, compasses, pistols etc., costing much less are at the other end. A vast majority of equipment lies in between with varying costs, complexity, population, criticality and utility. Obviously, we cannot have a common policy for maintenance. The policy will have to be tailor made to suit the type of equipment in order to ensure its desired level of availability. Equipment Management Policy defines the nature, depth and estimated frequency of repair to which the equipment is to be subject in service. It is based on the reports of the "Maintenance Advisory Groups" who are expected to work in conjunction with the Research and Development authorities whenever the equipment is either procured from abroad or developed indigenously. In the case of purchases ex-trade these groups subject the equipment to extensive tests and trials in order to come up with appropriate scales for spares, tools and workshop facilities. Since, maintainability also alludes to both the needs of the materials that are handled by the system, but in reality are a part of the system; and the needs of the system to arrive at maintenance concept we need to make a choice out of a diverse range of options available. To have a closer look at the options available and their effects, the following three aspects require examination:

- The behaviour of the object to be maintained, i.e. an equipment when it is introduced in service will have a reliability and maintainability through design and manufacturing, independent of the use of the maintaining organisation. For, in real situations these properties will be influenced by the way the objects are used, by their environment like deserts, high altitude, and plains etc., and by the capabilities of the maintaining organisation.

- Requirements of the operational commander, wishes and constraints from production agencies prescribe the level of reliability of an 'equipment'. In simple terms this means that reliability, and availability should have high values. Our 'echelonised' repair and maintenance structure will also govern the methods to be adopted for ensuring lower rates of the equipment failure and its being non-available for operational use.
- Simplicity in weapon and vehicle design provides for relatively easier production, operation and maintenance. It can also be achieved by cutting out superfluous appendages that generally are only "sale enhancing" gimmicks. Application of advanced technology and adoption of modular concept can considerably improve maintainability. Increased use of module units in electronic equipment has greatly increased their ability to be maintained relatively more easily. While selecting equipment a particular attention towards these aspects can improve availability of equipment by reducing its down time.

Maintainability has to be considered for all kinds of operational and functional situations. There is that famous case of the American M47 tank when it first appeared on the battle front. It was found that for changing spark plugs the fitters had to first remove the engine. There were many other flaws creating all kinds of problems where maintainability was concerned and it took a lot of time, expense and effort to rectify those. The user, in the meanwhile kept losing confidence in the product. It will pay rich dividends if we associate our maintaining agencies and the users right from the product development stage.

The need for simplicity in planning, in the design of supply systems, and in the maintenance and operation of equipment, if not its technology and functioning, goes without saying. It is a principle that has been followed more in the breach than the observance on more than one occasion.[8]

Standardisation

Today, a very wide and diverse range of military equipment in the form of transport and utilities in addition to munitions is in usage. In all, the inventory of the "Three Services" comprises approximately 12 Lac items. Besides causing a tremendous load in the form of spares and reserves which are required to be carried to the operational areas for a variety of vehicles and other equipment, lack of standardisation does not allow for a slim and trim inventory. This also means dedicated workshops and consequent waste of technically trained manpower that could be making meaningful contribution elsewhere. In the case of clothing items we have too many sizes. (There are as many as sixteen sizes for shirts and trousers). This often leads to faulty provision and procurement action. In our system even a slight change even in the packaging could mean a new part number thereby implying as if the item is newly introduced leading to a whole cycle of different procurement activity. (In order to visualise the magnitude of the task involved, as per the calculations of the US army, introduction of just one new 'equipment' causes a Divisional Ordnance Unit to in-scale at least 2000 new types of spares. It would need two four tonne trucks and one 10-tonne truck to carry these spares). Even amongst the three Services the types of equipment performing the same role and the task is vast. Amongst the 12 Lac items held there will be a good number which have identical or similar specifications but are known by different part numbers and names. This situation often results in excessive stocking in one Service with no outlet to the other, in respect of items that are common to all the Services. Thus, it is not unusual for one Service to be disposing off as surplus the stocks at throw away prices that are in acute shortage in another, may be even in the same station. It is now being accepted that increased standardisation will make it easier to achieve an overall increase in our capabilities. In our armed forces an effort is being made to standardise the items by allotting them common NATO codes and by developing a common catalogue. This will help in establishment of a standard language of item identification for equipment received from different sources. Effective utilisation of assets by establishing interchangeability and substitutability between items can be ensured. Effective coordination in procurement by helping to eliminate concurrent buying of the same item will be a major gain.

The standardisation exercise, inter alia; will have to include the relevant trade and industry as well. In the absence of a combined effort in this direction there is a risk of rival manufacturers plugging in for their products, thereby leading to some kind of dilution of the whole process. In the case of those items that are in common usage like vehicle spares and assemblies it will make better sense to allot code numbers by a central agency. At present the manufacturer allots his code number to an item, the three services give their own number to the same item. A sort of directory is brought out by more enterprising manufacturers where relative identity is established. Time, energy and effort expended in this exercise has not yet been quantified, but I have adequate experience of the confusion caused in the absence of standardised products and specifications. A central agency should undertake this exercise and the "standardisation cell" working under the ministry be part of this agency. There will also be need for `coordination and control' at the highest level for very valid reasons and also to ensure that the trade off between the defence and the civil usage does not in any way lower the quality of life of our population. There may be some compromise being made while fixing standards and identifying potential manufacturers who could be having sufficient production capability as well as infrastructure to have their products delivered all over the likely theaters where military operates. Upper end civilian users of fancy products are not however, likely to be affected because military as a principle buys an acceptable product at an affordable price instead of going in for items where cost is no consideration.

Bureaucratic structure and culture, involvement of too many ministries, general apathy of the production agencies and the user's reluctance to change tend to slow down the standardisation process. Therefore, if we have to (and we should) blend the defence requirements both for war and peace time with the civil end usage standardisation will be essential. This will also be important if an "Integrated Logistics System" is to be developed for the three services. The advantage and economics benefits of such a development are obvious. The subject of "integrated logistics support" with all the Three Services has been broadly discussed elsewhere. In the instant case it would be sufficient to say that the real estate occupied by the logistics installations of the Three Services, the manpower employed for the purpose of manning and maintaining the stores is high. If, some how, the three could be integrated without sacrificing efficiency in any way the financial benefits to the nation will be of a considerable magnitude.

Integrability and Interdependence

Integrability basically implies that new inductions and technology should, as far as possible, fit into existing holdings, training and maintenance support. Various types of equipments be as compatible as possible (preferably including Air force and Navy) so that instead of different agencies trying to maintain separate inventory for even small and common usage items, there can be only one agency dealing with it. This will be particularly useful if we are importing a particular type of equipment, and maintenance spares are also required to be bought along with the main equipment. If compatibility is established before hand one can be selective about what to buy. While buying a weapon system(or any other system) we should see to its integration in our environment (both strategical and tactical) which may not necessarily be the same as that of the country of its origin. For example, the BOFORS gun had primarily been designed for the NATO artillery and its over 600 kgs ammunition pellets are to be lifted and loaded in a support vehicle (Scania) which has a boom of over 2 meters swinging in and out of the vehicle. In our climate the loading mechanism also needs 15 minutes for cooling after half an hour of usage. We neither have the storage sheds which can provide sufficient space for the vehicle to be effectively used with its boom swinging in and out, nor adequate road network in our forward (hilly) areas where these guns are expected to be deployed. Tanks of East European origin getting heated up after some movement in 'hot plains and deserts' is not uncommon. Integrability of a new system be ensured not only from tactical angle but the maintainability and administrative angles as well. If a particular system needs setting up of a total infrastructure then it may not prove to be economical in the long run.

A totally and radically different weapon or communications system could take a lot of time, energy and effort before giving complete confidence to the users. This is particularly so where major changes in our "maintenance support system" are either envisaged or they come about due to continuation of training methods of an earlier system leading to speedy wear and tear. (It is not unusual for an artillery regiment to be carrying out their 'gun drills' developed in the times of 25pdrs even when equipped with a gun which was far too advanced with its sophisticated computer controlled firing and mechanised loading mechanism. When the author brought to the notice of the commanding officer the wear and tear being caused to the 'firing mechanism', he expressed his inability to do any thing about it because no new training pamphlet had been received and he could not, on his own, make any changes.) Similarly, management of palletised high caliber ammunition has been a major problem, more so in the forward and high altitude areas where neither appropriate transit and storage facilities are available, nor can these be created in a hurry.

Interdependence

Interdependence means that each part of the logistics system must be coordinated with the rest of the system as a whole. In this, one factor which is always likely to cause problem is the 'empire building' effort consequent to the induction of a new weapon system. This happened when missiles came in for the first time. The issue of their inspection and maintenance has not been resolved (in the best interest of the weapon system) to date. Ordnance felt that the missiles being primarily ammunition be also inspected as such. The corps of EME on the other hand were successful in convincing the decision makers that with the presence of electronic components missiles automatic fell in their domain. Consequently, we have a fairly confusing, and to any rationally thinking commander, an ineffective system of ensuring their serviceability. In so far as their deployment has been concerned we have had too many doctrinal changes ever since the missiles have been inducted. These changes seem to have largely been governed by a chance to increase the strength of a particular arm or service. Integration of a weapon system in such a situation becomes virtually impossible because overall organisational interest tends to get in the background .

In times to come RSI (rationalisation, standardisation and interoperability) will play an increasingly important role in the Military's future as the defence budget declines and commitment of the Services increases primarily due to proliferation of dissidence and violence in our border states. No harm to start considering these aspects now and plan our fresh acquisitions in a manner that ensures these three inputs.

As a first step we can form a committee consisting of members from all the three Services to: firstly, identify those current equipments where we can apply RSI; secondly, all fresh acquisitions in respect of weapon and communication systems be put through similar tests. If required "inter service QR" should be formulated and where necessary, in the interest of commonality a particular Service may even accept some dilutions of its expectations, provided these expectations do not relate to operational considerations. One factor that may have an impact on integrability could be the financial impact of the whole transaction. That is, the mode, method and cycle of paying for some equipment could override other considerations. It would be pertinent to mention that we have equipment pertaining to both Eastern and Western countries. With a little effort it should be possible to find common grounds with something already existing in our inventory. Establishing of such parameters will lead to economy, both in terms of money and maintenance facilities. The implication here is that unless it is some new, revolutionary and totally different technology it should not be difficult to adhere to this principle.

Information

Accurate and up to date information is vital to effective logistics planning and to supply distribution. It is as important to know where supplies are as it is to have them physically present. The "operational commander" can give an appropriate shape to his plans only if he has accurate information with regard to the logistics availability at the time and place where he wants and when he requires them. Similarly, the "support-service commander" needs to know as to where and when and in what shape the allocated resources are going to arrive. He needs this information to plan for the space, manpower, technical complement, security etc. He has to ensure that he provides required means for a successful implementation of the operational plan. With the arrival of computers and improved means of tele-communication it is possible to keep a total track of the stores in the pipeline.

In today's environment of computer based information systems it is possible to have accurate and timely inputs with regards to logistics. A suitable 'Management Information System' can be designed to meet all the needs. Aim of this MIS would be to assist commanders in decision making and to provide decision support system for various classes and types of logistics related problems .Information should be available at various hierarchical levels in the chain of both the operational and logistics channels so that commanders at all levels are aware of the availability or otherwise of their material support. Information, of course has to be timely and current for efficient and correct decision-making. We may also add "logistics intelligence" under the head of "information". This intelligence should pertain to the likely enemy's capability, as well as capability of own organisational means and the industry to fill the gap between a commander's needs and availability. Logistics intelligence with regard to capability and availability of local resources to support an operation will be a big help while carrying out advance planning. Alexander the Great could burn his wagons and chariots carrying material support because he had learnt that the areas he was to attack in the immediate future could meet his requirements of material for successful conduct of his operations.

Some of these principles were evolved during the 2nd World War, and during Korean and Vietnam wars. Kuwait operation has added another aspect to the principles of logistics. As these are examined along with the problems and issues involved it has lead to a situation which has brought in yet another important principle-"Generative Logistics". This principle consists of application of initiative, technical knowledge, and ingenuity to accomplish the mission. The importance of this principle is that it is difficult to predict all possible events in war. Units can overcome friction, time, and distance factors by applying the sustainment imperatives -

integration, anticipation, continuity, responsiveness and improvisation. If the theme is generative logistics then responsiveness must be ensured. If the theme is economy, then the processes will focus on frugality. If the theme is efficiency, then the processes generate "cost effective" solutions. Cost-effectiveness rationales are the most analytical in their themes, tending to give minimum consideration to human nature. Total mission responsiveness includes considering on the move repair and associated human skills and maintenance equipment requirements. The support tail determines the ability of the combat force to continue and sustain its mission. By addressing the total mission profile, we should begin to see how to create a logistics capability responsive to the overall combat needs. It is never too late to accept that under most conditions, logistics responsiveness will enhance the effectiveness of a combat system, provided the logistician is a part of the team that defines the strategy, and designs tactics to achieve those strategical goals.

In fast moving, fluid and intense operations the rate of attrition of both men and material is colossal. Logistics involve hourly and daily problems of organisation and movement. To make matters even worse nobody really knows what a modern high intensity conflict between two relatively balanced opponents might require in the way of logistic support. Therefore, a serious problem for the logistics planner is determining whether the next conflict will be short or long. Our past experience indicates short and intense conflict. Even though there are certain gray areas in so far as logistics support for a short and intense war is concerned there are also pointers. The Israeli/Arab conflicts, our own short intense wars since independence and the 1991 liberation of Kuwait all give some indication of what might be needed. All of these could be considered to have had special circumstances which, while not invalidating any logistic lessons, does mean that their applicability to a modern high intensity conflict between two relatively balanced opponents could be questionable, particularly when we have been engaged in a long drawn "low intensity conflict" for over ten years now. Our troops are deployed for indeterminate periods in "internal security" duties and are too far spread out for the integral logistics units to provide meaningful and assured material support. This means that our traditional methods of support have limited utility when the troops are given an unconventional role with no specified time frame or likely duration for their deployment. There is an urgent call to reexamine and redefine the role of logistics units keeping the ground reality in perspective. An important aspect that we should look into is that the material management in the army has become very complex in magnitude and quantum due to environmental factors. Unfortunately, we are still giving no particular consideration to logistics assessment, deployment and replenishment aspects in a manner that could provide a higher level of assurance to the combat soldier.

Another cause for concern is that often enough the commanders do not seem to appreciate the fact that like combat units the logistic units also move by 'bounds'. That is to say, if and when combat troops advance, the logistic units will also have to advance along side. Only difference is that, for the combat units the 'bounds' are tactical and usually relate to the next hill feature or a valley, or may be a piece of high ground. For a logistic unit it will have to be an area where a workshop or a field ambulance or ammunition refilling point (all of them very lucrative targets in their own right) could be set up. To compound the problem, these support units, due to their size in terms of weight; volume and nature of war like stores are, perforce tied to the existing road or communication net work. Because all the assets in hand makes these Service units attractive targets they need far greater amount of camouflage and concealment. Consequently, these units are relatively slow in keeping pace with the supported troops, and unless rehearsed well in advance and provided with commensurate mobility can slow down the implementation of the tactical plans. This could lead to situations where commanders will either have to make compromises or supplement their logistic units, an activity not easy to undertake during actual hostilities. To get an upper hand, retain and sustain the initiative action required at all levels demands prompt, efficient, and responsive replenishments of the wastages accrued. In an actual war, and not appropriately brought out in exercises, such considerations can have a serious impact on the ultimate success or failure of the mission of the combat troops, as the Americans learnt (and as combat literature tells us) gradually and the hard way in Vietnam. In due course of time and after having lost considerable number of men they realised a very simple principle of logistics. They found that in Vietnam that their logistics planning was complicated by the fact that logistic troops and units were deployed at about the same rate as tactical forces rather than in advance of them, as desired for the timely establishment of an adequate logistic base because of the time and effort involved.[9]

For an army to be efficient and always ready the equipment has to be kept in working condition, the troops have to be rationed, kept healthy and happy. The officers and men responsible should be involved in all aspects of the profession. Only then a meaningful and economical support can be provided to the dependent troops. We have established an 'Operational Logistics Directorate' to coordinate and ensure timely placement of resources in respect of the "field forces". Some officers and staff from various Services have been included to interact with their respective Service, but they hardly have any say in the planning and actual assessment of the logistics effort required in a particular theater. The 'Service Heads' are only required to react to the demands of this ' Directorate'. In this we seem to have overlooked the fact that the evaluation of logistics effectiveness is one that needs the finest kind of mature, well defined, professional and fully informed judgment from experts in different areas. This needs a lot of specialised expertise in a vast array of fields of inventory and resource management, technological, medical and technical skills, considerable real estate and manpower. With ever growing technological advancements, with newer and more lethal and, therefore more intricate equipment coming into the hands of troops, the need for the support services to possess expertise in an ever widening range keeps growing. Furthermore, armament technology is moving at a remarkably fast pace and new weapons and equipment are constantly being introduced in response to the tactical lessons learnt from earlier conflicts. They all have a more complex and intricate impact on logistic planning. To achieve all this and much more the services have various branches providing material support to the combat forces. Rather than making full use of this expert body we have assigned this task to those who, at best, have a nodding acquaintance with the art and the science of logistics. Logistics is all about organisation and it is no armchair job. It is not an area where amateurs and the use of superficial statistics can contribute meaningfully to our national security. What we have to understand is that Logistics is a distinctive branch of warfare, and that it embraces a large number of activities that should only be coordinated, but not confused, with tactical or strategical activities.

Doctrine may not always apply in response to a given situation. However, by anticipating requirements and actions before, during, and after a battle, we can overcome the unexpected difficulties. Logisticians must be flexible and able to respond to any battle tempo and the commander's intent. Adherence to the principles discussed above will help him in doing so. Security has not been examined as an independent principle. Nevertheless, with the arrival of precision guided missiles our logistics installations are always vulnerable. We also need to conceal our logistics intentions from the enemy. He can always guess our intentions by interpreting our movement of material, a movement that can hardly ever be concealed. Similarly, principles like "Foresight", "Economy", "Simplicity", and "Cooperation" etc have also not been discussed because those form a common platform both for a tactical plan and a logistics plan. Specifically only those principles have been discussed that are relevant to providing an efficient material support to the troops in different types of operation. Foresight in placement of material and economy in its utelisation will largely depend on the higher commander's perception of an operational situation. So many factors and facets are involved in providing logistics to combat troops that there can hardly ever be a simple plan. The logisticians should organise there support in such a manner that the combat troops do not have to go from door to door in order to collect their essentials. A simple plan will logically provide guidelines for support services to plan ahead.

Notes and References

1. *Logistic Support-Vietnam Studies*, by Lt. Gen. Joseph M. Heiser, Jr.
2. *Logistics Management*, by James R.
3. A rake is a train comprising of a specific number of wagons.
4. It entails prepositioning of stores at the least vulnerable strategic points. These are well dispersed and any dislocation and disturbance of the system does not prejudice the supply system.
5. Donelly, Soviet Use of Military History for Operational Analysis, as quoted in *Logistics - The Life Blood of War.*
6. *Logistics - The Life Blood of War*, by Maj. Gen. Julian Thompson.
7. *The Lessons of Gulf War,* by Anthony H. Cordesman and Abraham R. Wagner.
8. Major General Julian Thompson, *The Life Blood of War - Logistics in Armed Conflict.*
9. *Introduction to Pure Logistics*, by Stanley L. Falk.

4

Planning Aspects- Logistics

> A military maxim has it that amateurs talk about strategy while professional talk about logistics.
>
> Time magazine- August 20, 1990

Logistics planning and Planning for logistics

Basic to all logistics planning is the art of asking the right questions. In suggesting that the primary aim of logistics is " to get there with the most"(and the best), the question may be asked: Where is "there"? The answer probably would be the critical point. And where is the critical point? Very likely it will relate to logistics- a major river, a railway centre, a road junction, an industrial area. It simply is a case of mingling logistics with the strategic and tactical considerations in the basic decisions of war- choosing objectives, deciding where and when to do battle, assessing the cost of continuing battle, to reaffirm that the material support is available within the acceptable "turn around" time frame. Logistics planning can be used to indicate the incorporation of logistics considerations into the formulation of 'strategic and tactical' plans. In terms of the formal process of military decision "logistics planning" can be considered as the logistics aspect of the commander's estimate of the situation. It includes the determination of the basic logistics requirements and the general dispositions necessary to support these plans. All of this can be effectively done only by a constant and intimate relationship between the strategic planner and the logistics planner. In this relationship the former is constantly aware of the logistics capabilities and limitations, and the latter is constantly aware of the dispositions and employment being proposed for the forces and the estimated nature and strength of enemy opposition. The likely duration of an impending operation and the capability to sustain troops for that period and beyond, if required, will provide relevant data for making support plans. An analysis of the assets versus requirements will tell the operational commander whether his plan is supportable or not. The relative importance of a particular operation to overall success of the strategic mission will dictate to higher policy makers to decide on either making good the deficiency or to altering/modifying the plan.

There are two main components of a nation's security policy- economic wherewithal and threat perception. Strategy and national objectives form the core of the whole spectrum and logistics start with the fact that much of the strategy grows out of the economic situation. Logistics at the strategic level means providing the force at the place where national decision-makers decide to use it. And logistics at the operational level are the ways and means of supporting a force at the place of military intervention. If the nation's industrial, technological and economic capabilities do not support these plans then they are meaningless exercises. If the nation is dependent on foreign suppliers for most of its requirements then also the plans have only a limited relevance, because one does not know when that friendly foreign supplier changes his mind, or finds a more promising customer. Therefore, the nation's capability to maintain and sustain a strategy must guide the planners. One has to put his perception of the strategic situation on paper so that what is required to achieve those ends can be properly analysed. If the strategic and logistic plans are to be brought into timely harmony they must be fully "integrated" from their inception through their final execution. The requirements of strategy should determine the composition and employment of forces and their associated weapons and weapon systems, and other ancillaries that are needed to support the strategy. These include the communication networks, technically trained manpower, industrial, and technological bases in diverse fields, encompassing almost every area of governmental activity. This implies a 'set up' to monitor and control all those areas that augment the defence and assist in maintaining forces in a viable, practical and economical manner. That is to say, for carrying out the approved national strategy, the forces to be maintained and their roles and missions should be so designed that it is possible to exercise appropriate control in each area in which control is required.

In so far as the logistician is concerned, he is interested in the design of the whole machine, because a large part of the preparation is in the field of logistics. He must know the intricacies and relative importance of various tactical options so that he can plan placement of the logistics elements accordingly. The qualitative and quantitative analysis of the logistical task must be based on the nature and extent of the commitment of the whole formation. Before implementing any plan for logistics support activity, the logistician must ensure that the tactical commander is allowed to concentrate on his primary mission so that a clear perception of the whole thing is available for planning and procurement of resources. Logisticians should also indicate the likely time frame required for making preparations complete. There are many activities like conveyance by rail up to the railhead, transshipment, matching up of the ammunition items with particular types of guns deployed on different axes, obtaining appropriate oils and greases for weapon systems etc that go into providing a planned support. Since, a variety of agencies are involved, coordination of activities could lead to delays and uncertainty. To safe guard against such a situation, and to minimise their impact both the logistician and operational commander have to be very clear about what they want, where they want and when they want. This is important because once placement of resources start in a particular manner and direction then it is extremely difficult to retrace if there is any major change in "tactical thinking". The operational commander should know that the required (and approved by the highest commander in the theater) logistics support would be available when ever and wherever required and in a measure that is essential for his operational task. To achieve this end the logistician in concert with the operational commander must analyse to see whether the logistical plans are based upon the strategic plans and tactical concepts of the higher commander. Would he be able to match the needs of the most probable courses of action in times of war, should be his constant focus. For this he needs to be taken into confidence at the stage when strategic and tactical plans are being made because the operational logistic action is related to specific strategic and tactical plans. In the absence of such an active coordination there could always be a mismatch between what is required and what can be made available. This suggests the 'equivalence' of strategy, tactics and logistics. Certain thinkers have advocated 'interdependence' that is described as the interrelationship among all functions. In their opinion the interrelationship among strategy, tactics and logistics ultimately drives the course of action taken in a contingency; and anticipated actions must blend with the strategic, tactical and logistics factors. We know (but do not often want to admit) that strategy, tactics and logistics are different aspects of the same thing. If separated they become meaningless. Unfortunately, logistics still remain secondary in some thought processes, ignoring the fact that if any military plan is to be realistic it must start with consideration of logistics. (A laudable exception where logistics considerations overrode political, strategical and tactical considerations is the 1971 war when the then Chief of Army Staff Field Marshal SHFM Manekshaw insisted that he must have his logistics elements built up to the desired level before venturing into any operation how so ever desperate the political situation otherwise was. The gaps in our logistics elements which could hinder the strategic and tactical aims were identified and filled to the extent possible employing all available resources before he told the government that now he was ready for carrying out the mission assigned to the military. It took him almost nine months to have resources in place but he could lead India to a decisive victory inside of fourteen days because he was convinced that without material support no strategical plan could guarantee success.) A clear lesson from this case is that logistics considerations and plans must be interwoven with national, strategic, and tactical plans at all levels of command. Because the operational logistics actions are related to specific strategic plans, the logistics capability to support such plans must be the take off point while examining strategic options. Commanders who learnt their lessons well have stipulated, " first with the most". Any lowering of the importance of logistics in relation to combat functions will not allow this aim to be achieved. Therefore, proper integration of support structures and concepts into the commander's tactical plan is critical to success on the battlefield. This integration is only possible if the support services are made privy to the operational plans in the initial stage itself. Often this does not happen.

The process of "planning for logistics" ultimately determines the flexibility and sustainability of the combat forces. It will include details with regard to availability of resources, plan for the logistics build up, and identification of those 'base and theater depots' designated to support a particular operation. Among other things, it involves the details of supply, the build up, the cargo movements and the provision of medical and repair facilities. This also includes industrial capability, both available and that which can be created over a specified period of time. There are, thus three aspects of logistics support to the troops. These are (a) Strategic Logistics, (b) Operational Logistics, (c) integral or direct support logistics also known as Tactical Logistics. "Strategic Logistics" encompass the total national capability in terms of men, means of communication, material, industry and technology; and where the base depots, arsenals, and base repair facility etc are to be located for supporting operational plans. These are very deliberate decisions because creation of these facilities is expensive and once created it is virtually impossible to reverse decisions.

The **Operational Level Logistics** encompass the support of deployed forces on the scale of an Army corps. At the corps level, logistics planning aligns with specific theater-oriented operational plans using and deploying forces in a well-defined manner. To meet the "theater- based requirements, regional depots, workshops, medical and supply facilities are established more or less on permanent basis. To meet the requirements of the troops, "Operational Logistics" units, and regional depots where certain stores are pre-positioned, are established. In certain cases, and depending on the terrain, these also include the integral logistics units of the formations, even holding those heavier/ bulkier assemblies etc., which are cumbersome to convey when operations are on or are envisaged. Thus, "operational logistics" involve the zone of transition from "wholesale" to "retail" support. This effectively involves breaking of the 'bulk' loads, often a shift in the transportation mode, and making loads lighter by re-packing where the loads may have to be yet again transferred to either mules or made into 'man packs'. All these are time- effort- and manpower -consuming activities. In so far as an operational commander is concerned, managing this transition forms the focus of responsiveness at the operational level. Responsiveness is a linking factor and a measure of merit. Responsiveness must become the byword for logistics support planning, much more than a mere functional requirement. This stimulates answers to both the "how" and the "why" questions in so far the material support to the troops is concerned. This means that the logistical operational planners must respond with a tailored support scheme. They have to organise the distribution function in such a manner that the right things are delivered at the right places, over the most expeditious routes, using the least number of vehicles and facilities. More than anything else, at the corps level there is a requirement for adequate ability to gather, analyse, and synthesise information about the entire corps battlefield. This is necessary because in most of the cases corps is spread over very wide area and has divisions with "dual role" in all together different terrains. Creation of reserves for such dual tasked formations ties down resources in a big way, and need lot ingenuity on the part of logisticians. The main problem is that of holding 'ammunition reserves' at different locations because, besides occupying a lot of real estate the ammunition items have a prescribed "shelf-life". If not expended within that period, these items could lose their serviceability and become dud. It would be more appropriate to create suitable accommodation for ammunition in our base and regional ammunition depots and hold reserves in these locations to be moved at short notice to the areas where the designated formations and divisions move for operations. Another area of logistics concern is that due to operational reason brigades and below are deployed too far away from their integral support. In such a situation it is not unusual for support down to brigade level to be directed from a corps-level or even command level unit or depot. Such arrangements must be worked out much in advance. The troops should have flexibility to draw material from any convenient source. It may happen that unpredicted calls may cause some problems initially, but it should be possible for the service units to meet these demands out of their reserve stocks and seek replenishment from the base, or have the stocks transferred from another installation within the command. This is a technical control function and the service head at command headquarter will require up to date data with regard to those scarce and essential items that have a direct impact on operational readiness of the affected troops. He may have to create an " information management cell" for such a technical control function where records are regularly updated.

Tactical level logistics are held at divisional and lower levels. Since all units of the formation need this level of retail support, tactical logistics must take place throughout the divisional theater of responsibility. Tactical logistics are also the "first and second line" resources of the combat units. There is always a tendency to carry every item authorised in the "equipment tables" for 'just in case situations'. Very often this creates avoidable dead weight, particularly when a unit is heavily committed in an operationally fluid situation and finds its mobility restricted because of material in hand. A detailed plan, if made, to ensure that only bare necessities are carried in the first instance can lighten the load considerably. Better knowledge of how support works and how material needs, both seen and unforeseen, can be met will provide greater confidence to the troops. It can lead to establishing realistic requirements in the first instance itself. Combat elements should carry only what they need and what they should have. Then only discipline and self-confidence will control the temptation to carry an extra amount of fuel, food, and things like that. The remaining stores could always be conveyed under the aegis of higher head quarters once either a need is established or the situation stabilises. One thing to be guarded against is that our first and second line scales were fixed when the means of communication were really primitive in certain parts of our country where troops are committed. In the case of ammunition items due to higher caliber of weapon systems and faster rates of fire, scales as they exist may be inadequate in a large number of cases. So also will be the case with communications equipment. A revision of these scales, reworking of the "load tables" and fresh allocation of transport can be done, preferably under the aegis of " Operational Logistics Directorate". This directorate can adopt a mix of "top down" and "bottom up" approach so that more realistic scales are prescribed in consultation with all concerned.

Diagrammatically the planning for logistics where all the three ie strategic, operational, and tactical will be inter-woven can be represented in the following manner:

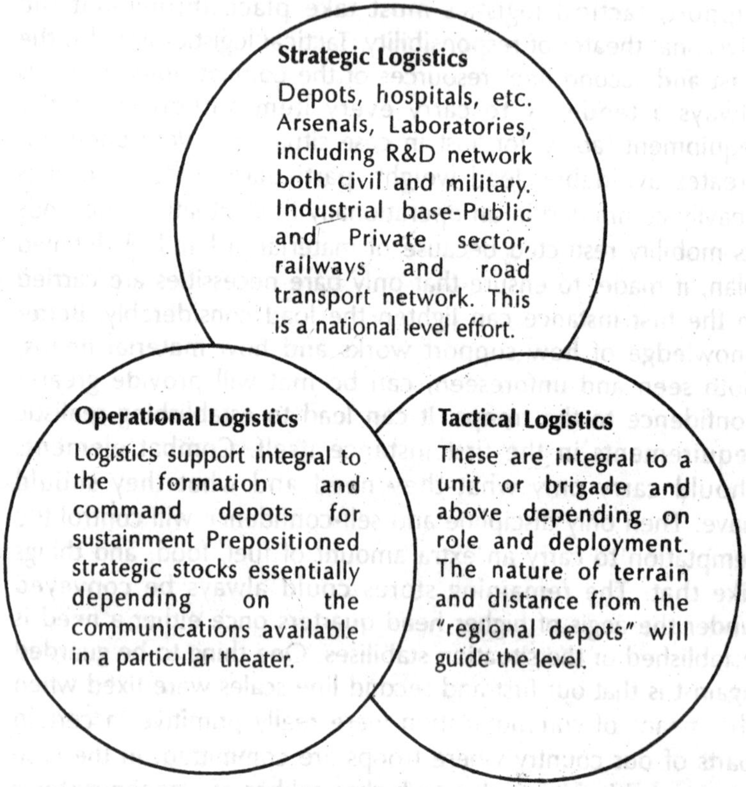

Primarily placement of the "Strategic Logistics" defines the operational strategy of a country. These placements generally are of a permanent nature and are not displaced unless the regional situation so demands. Very often units are provided direct support from the base depots when they are located in their nominated "peace locations". This helps in centralising those assemblies and spares and other assets like costly medical facilities etc, which are in short supply, or are expensive. Units and formations develop a working relationship with the depots that can come in handy during operations. Similarly, operational logistics units could also be providing cover to units that are not necessarily dependent on them. This is more so in the case of units and formations committed in IS duties. The situation may demand that such units and formations be provided support directly from these installations. Often enough a combination of strategic and operational logistics is employed depending on location of a particular type of combat unit and its operational role. Regional depots and integral support units are generally not geared to take on the additional loads because a perpetual shortage of fast moving essential items is a major constraint. This is not an insurmountable problem and an advance notice of some kind can make installation commanders take necessary provision actions. The difficulties will get compounded when a, non-dependent' unit arrives with equipment or vehicle fleet in whose respect maintenance spares etc are not held in the "area of supply". In such a situation the logistics units can organise urgent collections of stores from base depots and where possible, procure from the trade where items of military and civil end usage are concerned.

"Requirements" Planning and "Capabilities" Planning.

The whole process of equipping the forces normally begins with the identification of strategic goals and the country's response to its environment. It can be considered as the first stage where the basic problems of logistics need vis à vis strategic national goals can be identified. The means required to meet these goals also are taken into account so that availability of resources could be given a thought at an appropriate level. This stage is called "requirement planning". This process of requirement planning will follow through the sum of courses of action to indicate the tentative requirements of each military Service, and in turn the sum of all military requirements. Logistics mean heavy investment, both in terms of money and manpower. In order to achieve optimum results from our investment it is appropriate to analyse our requirements of material, spacing out of these materials inflow, and to vet these in relation to our strategic and operational plans. There are any number of reasons for undertaking the planning of requirements and the nation's capability to meet the security requirements. Primarily the reasons are: —

(a) **Judicious allocation of resources.** Resources are always limited. The government would want to deal with the situation in a manner that causes minimum inconvenience to the people. The exercise involves economic and industrial capability of the country. Prioratisation in terms of those sectors of national needs which are more and more demanding, and identification of those spheres that contribute to the defence ensures that the development efforts are concurrent to the defence requirements and preparedness. This may involve some special consideration where areas of defence interest with civil usage have been identified in advance. These considerations could be in the form of communications network, development of a particular industrial project or technology. Blending the military requirements with general public's well being helps in generating better facilities for the community.

(b) **Continuous changes In international power equations and consequent shifts in threat profiles.** The world has entered a very dynamic phase of fast-changing mutual relations. At the point when we evaluate our international relations there must be made the large strategic decisions as to which courses of action can be retained, or which must be discarded, or which must be reduced in scope. All this will be dependent on considerations of logistics feasibility within the context of the national policy at the time. And, of course, the country's capability to provide adequate wherewithal to the military so that the national goals can be met. A regular and periodic review of the situation, particularly of events in immediate neighbourhood will be of paramount importance in taking long term decisions.

(c) **Short and intense nature of future wars.** We cannot prepare for all kinds of war, but we can prepare for more than one. This is directly related to the capabilities of our potential enemy to employ his resources both in the form of military machines and his fighting capability. We must try to take into account the potential enemy's position, ability to sustain an operation and probable response to own preparedness. This will tell us whether our forces are strong enough, and properly designed in terms of equipment support in view of his situation and capabilities? Historical background of the past operations, his present and future capabilities will provide sufficient guidelines.

(d) **Long lead times.** This is a perennial bugbear of the planners. The administrative lead-time in itself, ranges anywhere from one to ten years before a decision could be considered to have been arrived at. Thereafter, the potential supplier may take anywhere from six months to four-five years in meeting the requirement. This indeterminate factor causes avoidable waste at all levels, besides creating a situation of uncertainty. We must first of all define the "administrative lead time" in a realistic manner, and monitor every important flow of files, and documents connected with defence requirements. Supplier's lead-time can be made more realistic and responsive by actively invoking 'penalty clauses' where and when required. Usually, defaulters are the PSUs and Ordnance Factories. The DGOS (Director General Ordnance Services) is a member of the "Ordnance Factory Board" but hardly has any say in their functioning. The problem and its possible solutions are being discussed in this narrative a little later. At this stage it is recommended for consideration that if efficiency and promptness are rewarded we may have a little better response from these agencies

(e) **Fast rate of obsolescence and technological development in the field of armament.** The equipment obsolescence rate is getting higher and higher with newer technologies emerging fairly regularly. This is particularly so in the field of communications, armament and ammunition. The problem is further compounded by the fact that technological forecasts are somewhat difficult to make. More so in a country like India where our own technological base is still in an embryonic stage and will take some time to mature to a level where our reliance on others decline. Another thing is that the fields where emergence of a new technology will change our way of planning and providing material support are virtually unpredictable. We can, nevertheless, always draw lessons from other countries and make our plans in such a manner that unless there is a major breakthrough in a particular field we need not bring in any radical changes in our logistics support system.

Note: All ammunition items and missiles have a prescribed "shelf-life". If the item is not consumed during this period chances of its becoming unserviceable cannot be ruled out. It is a typical catch22 situation. Reserves of ammunition have got to be held because things cannot be manufactured in a hurry. If we don't hold reserves we may be found wanting when trouble arises. If we build up prescribed scales of reserves, and if a major operation does not take place then we are left holding dud ammunition with its consequential problems. One way to guard against such a situation would be to specifically identify those items that have an extremely limited shelf life. These generally are artillery fuzes, and primers, primary and secondary cartridges for mortars, guidance systems for missiles and such like. Factories should specify the time frame they would require to manufacture quantities needed for making an artillery shell or a mortar bomb functional. Based on such an assessment only the "shell" and the cartridges be manufactured and stored in our depots. Complete rounds are held only for the interim period that is considered essential for responding to an emergency. This may involve some extra effort in the 'holding depots' but an overall economy will be assured. Another thing is that the arrival of palletised ammunition has created a need for "material handling equipment" on the battlefields. This has brought in a need for different storage and handling requirements even in places where the ammunition storage locations may not have proper sheds or tracks for the MHE (Material Handling Equipment- these are usually cranes, forklift trucks, pneumatic jacks etc) to operate. Dealing with these kinds of situations requires a great deal of innovation and ingenuity on the part of tradesmen from the services. I have myself used small 'sledges' used by children in the hills for shifting heavy pallets (individually) from one location to another. Rope and Pulley is another method employed for the purpose.

Deciding on Requirement Level

The strategy and the information with regards to logistics requirements of the prescribed military strength, coupled with the necessary information on where and when to be positioned can lead to arriving at approximate requirement levels. The controlling agencies could thereafter have the details with regard to precise capability of each source for meeting military needs. With the information thus available resource allocations can now be made to the Industry wherein the peacetime requirements can be met in an acceptable time frame. Identification of possible 'civil end' usage of a particular product can be a big motivating factor for the suppliers. The concerned industrial sector, be it private or public, can be given indications as to the likely operational requirements identifying priorities for supply. This will usually be governed by consideration of the 'terrain of operation, deployment of troops, and time and effort involved in having stores materialise in a particular theater. The lead-time that would be given in case of operational situation can be also indicated so that the suppliers could arrange their production cycle accordingly. This will help the suppliers plan their raw material and other commitments well in time. This could be done even in certain cases of imported equipment where we have a long-term relationship with suppliers, and a fair assessment of the wastage pattern is available.

Capability planning consists of the determination of what kind of forces you can create, employ and support with means that are or can be made available in accordance with the strategic and operational plans. Thus, this can be termed as the stage that delivers the finished goods. The process will involve a detailed examination of our economic, technological and industrial capability to meet the projected requirements. It will also examine the repair, overhaul and maintenance capability. Wherever there is a mismatch appropriate requirements will be projected and provisioned for in order to ensure the desired capability level. Assuming that every thing becomes available the task of conveying the material to desired locations with dispatch will be a major consideration. This means that transportation in itself is a major factor where capability and capability planning are concerned. Judgment as well as analysis must play an important role in assessing the nation's strength and deciding upon the share of national resources to be devoted to defence. In this age of high speed computers, various factors and methods used to assess the military needs should be susceptible to rapid data inputs and mathematical manipulations in order that a large number of situations and options can be readily and speedily evaluated. Both kinds of planning (requirement and capability) takes place at the national level. It involves the finance, industry, defence, science and technology departments. Because the military does not exist in isolation only a certain amount of capability can be assigned to support troops. Where the need is critical we can lay down short and long term targets for the facilities to be made available.

It is to be remembered that all planning factors are based on experience or usage data, some of it good, some of it very poor, some of it up to date. Therefore, these factors do not become good planning tools unless the planner knows the circumstances under which the fundamental usage data on which the factors are based were collected and processed. Important questions like how to make resources available with minimum strain on the national economy will be thrown up when the computers generate a 'decision matrix'. These are questions that must be considered and debated at the highest level of policy and planning. The process in any case does involve some of the same people, problems and component organisations. Can it therefore be integrated so as to derive optimum benefits for the society, is an aspect that needs to be examined at length so that the national defence and economic growth could be better intertwined. It is also pertinent to stipulate that experience would have already indicated the critical areas of inadequacies that would need governmental level action to ensure that the national goals and objectives are met. In a country like India there could always be political compulsions of the government that will tend to dictate decisions that might not be actually the best for the country. In our environment where every requirement has to be justified and often, in fact, has to be lobbied for years there will always be a clash of interest between the perceived and actual needs of the Services. (The press has been talking of bulletproof jackets awaiting bureaucratic sanction for over 8 years, with men being killed in counter insurgency operations). This will require the service chiefs to ensure that the national interests are not compromised for political mileage by the government of the day, or bureaucratic insomnia where the man sanctioning a military requirement has neither any soldiering experience nor concern for the safety of our soldiers.

The ideal situation for an Army logistics organisation would be to improve its operations to the point that the only constraint is the availability of funds for a particular requirement. That is, budgetary constraints should be the main guiding factors while planning our procurements. This would mean that the organisational system should become fully effective and be able to meet all demands placed on it In other words, the system as a whole would have excess capacity. All (reasonable) customer demands would be met completely and on time by the organisation, and the excess capacity could translate into a wartime surge capability with a possible civil end usage when military demands are limited. This also means that given today's shrinking military budget and ever increasing and, often conflicting, demands placed on it there is an urgency to manage the ' Support Services' as business enterprises. There would be a renewed interest in efficiently reducing inventory investment. In achieving this reduction, Army inventory managers must carefully weigh the efficiency gained in the shift to the maintenance philosophy against 'war wastages reserves' and contingency sustainment. All that this implies is that the objective of all logistic effort should be the creation of continued and effective support of the combat forces. And the combat commanders, in turn, must make logistically supportable plans. This they can only do if they take their logisticians in confidence, and do so well in time, so that resources could be marshalled in a manner that provides the best solution in a given situation.

In developing an appropriate national-security posture and the allied considerations for material support, we should begin with a doctrine and work towards force structure and requirement. Rather than begin with equipment and force structure and then work towards a doctrine (as is often the case when the resources are limited and priorities are conflicting). Certainly this is not easy, a lot of hard thinking by the government and particularly the planners will be required. Proper planning requires - a defined strategy, a long-term affordability constraint, and the constant iteration of the two in order to arrive at an affordable and supportable strategical plan. We must also recognise the fact that, for many years to come, the potential strategies will be largely limited to those that can be achieved with the resources already available and the resources that can be created by a judicious mix of all the factors and considerations discussed above. With the induction of newer technologies and with changes in the international environment relative merit of these factors is apt to alter. Some fresh input may also necessitate additions and deletions to the list.

5

Defence and Development

Very knowledgeable experts talk of 'Defence vs. Development. Even a war-veteran like President Eisenhower of the US had raised the issue of the military-industrial complex. He had asked how many schools could be set up for the price of a 'fighter aircraft', or how many hospitals or miles of road constructed for the price of an aircraft carrier or submarine. In our own country itself, some well-meaning and powerful individuals always plead for, what they consider, more pressing needs of the country. In their zeal to improve the lot of the common man they insist on selling the theory that the soldiers always, and without sufficient reason, create an atmosphere of fear. And therefore, instead of spending money on building up the country's potential for efficient defence the government of the day needs to spend more on upliftment of the poor. Proponents of cuts in defence expenditure propagate that it has grown unabated, and with limited resources available for development the country has paid a heavy price in the form of 'opportunity costs'. This mind set has come about because over the years a view got created that defence and development are mutually exclusive and you can have one only at the expense of the other. The origin of this theory is not clear because the "socialistic" pattern of industrial and economic development that we adopted on independence makes no such stipulation. Russia, China etc., after whom we formulated our industrial policies have always treated defence as their 'core objective' while adopting any expensive development project. The defence production units are important elements of national industry in both the advanced and erstwhile 'East bloc' countries. Lack of a total view of the global environment and its impact on our national security has lead to a situation where defence and development policies have been totally disconnected. Consequently, compartmentalisation and the absence of such an integrated view have adversely affected both defence and development. Neither at the political nor at the top bureaucratic level has there been a policy-oriented approach to the security and developmental problems of the country.

United States and other leading arms producing countries view their defence industry as the spearhead of advanced technologies that provide valuable spin-offs from which all sectors of industry could benefit. They consider the defence industry not only an important segment of national industry but as its main thrust line for introducing new technologies. The future technological, rapid transportation, laser and other fields are studied to derive their implications for the future security needs of the nation concerned. This telescoping of the two allows their defence sector to pay for itself by generating more jobs, more trade in the international market and by generally creating a technological spiral that has helped in improving quality of a citizen's life. That the defence industrial base is a valuable resource for a nation's security and for its economic competitiveness hasn't as yet been absorbed to an extent where we take urgent action to update our thinking. Our philosophy of compartmentalising the defence industry from mainstream has not allowed us to fully utilise the facilities that are available in various spheres. We have been slow to realise that defence related industry and technology, in addition to everything else, helps 'savings' in foreign exchange, and gives the industry and the technology a boost.

Pandit Nehru had emphasised, in his equation of defence, "Your defence forces plus your industrial background plus thirdly the economy of your country and fourthly, the spirit of the people are the main ingredients". It is rather surprising that our first Prime Minister who had a vision on the issue of national defence that he put across so succinctly, did not give any specific direction to the policy makers to put his philosophy into action. This may be so because immediately after independence the priorities were identified in those areas that did not necessarily include 'self-reliance' in the field of military equipment and a focus on the national defence as a 'key result area' while formulating the industrial policy or other developmental plans. To some extent our defence industry or defence related civil industry, thus, suffered by default. Treating defence as a 'non plan' expenditure did not help the situation either because this allowed both political and bureaucratic ignorance on the subject to manifest itself in a manner which has put us in this situation of excessive dependence on others because we shied away from treating defence as the major thrust area for technological advancement and industrial growth. In the perception of our policy makers defence is non- productive in so far as development is concerned. These, so called, experts have tended to ignore the benefits to the society that could have been had if defence was considered as an important input for development. This compartmentalisation of defence and development, which started only as a convenient classification of expenditure, became rigid in the minds of not only the politicians and administrators but also of the economists. As a result, we have been rather slow in realising that the reason why many countries, including a few developing ones, have been able to maintain a high level of defence expenditure along with economic growth is because they have followed a conscious policy of amalgamating their commercial industrial sector with the defence industry and have understood that defence oriented development projects provide a sense of urgency to all concerned. They have had technological advancements primarily because they could identify the defence potential of emerging technologies.

It can be argued that the defence sector has not been given a chance to properly manifest itself because of our understanding of the prevalent politico-economic climate at the time of freedom. The world had just come out of a great war. Our leaders were more concerned with eradication of poverty, rehabilitation of refugees and creation of a welfare oriented state. The heated environment of 'cold war' also did not permit a defence orientation where our industrial and economic policies were concerned. We preferred to keep these two apart as giving a 'defence bias' to them immediately after independence could have affected our stand on world peace etc. Such insulatory tendencies have had an adverse effect; both on our defence related industry and development which could have been had in all the fields provided we had taken a long term view of the two, and realised that defence orientation of our industry and technology was the key to development. Resultantly, we are in a situation where not much headway has been made in arriving at a well-defined defence-linked industrial policy. We have, thus, been in a 'no win' situation, where slow pace of development has been attributed to the increase in defence spending, and reluctance to invest on defence -related research and industry has been ascribed to urgency for bringing in higher level of development. (After the military defeat at the hands of Chinese in 1962 our preoccupation with development was dished out as one of the reasons for our inadequate defence preparedness).

Having elected to have a 'welfare oriented profile ' and with fixed ideas about plan and non plan sectors(defence is a non plan sector) whole generations of leaders, bureaucrats, academicians et al have preferred to keep the defence expenditure as low as they possibly could. This has been done on the premise that defence expenditure restrains the country from undertaking development-related projects. Even military men have humoured these power centres by not making enough and forceful noise about inadequacies in our preparedness. They have also not forcefully focussed on the fact that by incorporating the national industrial and technological capability as a part of the Army's logistics strategy we could actually have had all round development as an offshoot of defence preparedness. All that was required to be done was, for the decision-makers to carefully analyse the political and military situation and, encourage industrial participation in providing goods and services to the military.

An important question can however arise. Should the defence requirements be an instrument of the country's 'industrial policy'? Are defence preparedness and the military's procurement policy required to be aimed at the efficient purchasing of equipment for national defence? Or, should the pursuit of wider economic and social objectives govern our technological and industrial charter? In answer to all these questions one can say that defence really helps the national economy in a big way is all that needs to be understood. It needs to be treated as an investment in national growth. All we are required to do is, (a) identify technologies that have commonality with civil usage, (b) Identify segments of industry that have a commonality, (c) ensure that things like area development plans in the border and remote areas are dovetailed with the border road development. In the US the entire highways development in the fifties was defence led. It is understood that at the height of the cold war even if a highway was to be laid, the aspect of its usage for operations was invariably examined and if required, appropriate changes were made to suit defence needs. This helped in two ways. Firstly, the civil population got what it needed in the form of improvement in communication and the country did not have to duplicate the expenses because there was a defence need as well. A simple lesson is that an improved rail, road network or better radio or telecommunication setup provides more advanced service to the citizens in addition to making itself available for defence needs. Secondly, and more importantly, the common man had no specific reason to resent what, ostensibly, was a defence expenditure.

Industrial Development and Defence

On independence, we went in for what was described as the 'public sector' with the aim of providing means of acquiring new technologies and acting as a growth for industrialisation, as well as contributing to import saving and possibly earning "valuable" foreign exchange. Under the First Industrial Policy Resolution of 1948, country's defence industrial capability was reserved for the 'public sector'. Inspite of technological inadequacies the existing Ordnance factories were expected to form nucleus for providing high quality equipment to our military. For a variety of reasons this did not happen. We continued to rely on imports in so far as our basic needs of armament were concerned. It is only after the 1965 war with Pakistan and the experience of the embargo by both UK and USA; the Defence Plan was aligned to run concurrently with the "Five Year Development Plan". A major objective was that by 1973-74 the country would be significantly less dependent upon arms imports. This policy was sustained in each successive five-year plan and today, there are around 39 Ordnance Factories and 8 Defence oriented Public Sector Undertakings (PSUs). These factories have been producing a fairly wide range of defence equipment for the military. They performed reasonably well during the late 1960s, especially in the production of arms, ammunition and vehicles. Notwithstanding the fact that their products were costlier and inferior to imports, the military accepted the situation in the fond hope that eventually quality will improve. This did not happen. Lack of technological upgradation, absence of competition, and an assured customer has brought most of these factories to a stage where their very existence is being questioned. To make matters worse, their performance in the fields of low technology items like clothing and general stores sections has been equally lack-luster, unpredictable and much more highly priced than the civil industry.

Defence Production Units - Historical Background

Defence production was first started in 1801 when the East India Company established the Gun Carriage Agency outside Calcutta. In the interim and up to the 2nd World War, the British established 16 Ordnance Factories. On independence we inherited this large number of 'Ordnance Factories' which gave us a useful base upon which to build on. These factories were originally set up (mostly) during the 2nd World War to meet the requirements of the 'Eastern' theatre and that too for a limited range of items. It appears that the British were not very sure of continuation of the Empire, and thus were not particularly keen on investing more than the barest minimum, be it technology, machinery, or expertise. At that time our national leaders did not have any say in the matter. The industrialists also did not have enough resources in terms of technology or expertise in a variety of defence-related equipments so as to make any meaningful contribution. Consequently, the bulk of production by these factories was and continues to be of low and medium technology even fifty years after independence because with a "captive military market" no need for technological upgrades was felt . It is only now with the end of the era of controls and rigid rules on patronising only the Public Sector Undertakings, the situation seems to be improving. But a large number of counter productive controls still exist. Another problem area is, that the management of some of the existing PSUs is generally not professional and competent. The fact that these factories require to be managed in a business like and professional manner apparently does not seem to have influenced our thinking in any particularly constructive direction. This lack of business like approach and our acceptance of the situation has not helped us in making our defence production units cost effective.

Today, the inherent weaknesses of trade unionism, a politicised work-culture and lack of quality awareness have created a situation where most of these units have turned into white elephants with low capacity utelisation churning out inefficient products. Generally their non-adherence to supply schedules have made most of these factories an undependable source of supply. It is gradually being realised that some of these Ordnance Factories with the usual laid-back approach of almost all government concerns and lack of competitive environment and with the trade unions holding the management to ransom are neither efficient nor cost-effective. It has been established that functioning of defence units in isolation has largely been responsible for production delays and escalating costs leading to situations where we often have to resort to either imports or purchases ex trade even when our production units are already working. To compound the problem for the military, since the potential suppliers know that the purchase is necessitated by a domestic failure they raise the price. We, thus, suffer a dual loss. Firstly, these unprofessional PSUs leading to failures and unacceptable delays. Secondly, we have to buy even simple items like blankets, and clothing essentials on vendor prescribed conditions. In this generation of high technology, expensive equipment, rising weapon costs and increasingly constrained defence budgets, government cannot avoid but question the efficiency (or lack of it), and competitiveness of these PSUs. This aspect of the working of the Ordnance Factories has attracted the attention of the 'Estimates committee' and the 'Public Accounts Committee' of the Parliament. Some of the usual comments are:

- The percentage of achievement of targets in respect of ammunition ranged between 45.33 to 60 and in the case of 'B' vehicles between 33.33 to 66.66 percent.
- Inventory holdings of the 'Ordnance Factories' continues to go up. This means more locked up capital eventually adding to the costs passed on to the customer, which a resource-starved military could, gainfully, utilise elsewhere.
- Obsolete and surplus material worth hundreds of crores of rupees, which is no longer required by the

army, is lying in Ordnance Factories. This is one main reason for Ordnance Factory products to be costlier than their civilian counterparts.

- There have been frequent delays and consequently cost over runs in execution and completion of projects. Level of uncertainty is so high that the user is never sure of an ordered item becoming available by the agreed upon date. Another thing is that the factories expect military to absorb these cost escalations which come about due to their inefficiency in the first place.

  (To counter the above, the 'Department of Defence Production' pleads that " Ordnance Factories" have not been set up for purely commercial purposes. The underlying theme is that a 'welfare' element is also built into this. Therefore, it is contended that any elaborate system of monitoring man-hour and machine-hour utilisation in these factories is not needed," and that "indirect devices" being followed are adequate for the management control). This stipulation however, is not likely to measure up to any economical or logical yardstick. No country in the world treats its military requirements with less than optimum priority; socio-political reasons do not interfere with military preparedness in other instances. In any case, one fails to understand why we cannot make these factories commercially viable and dependable when the infrastructure exists. It is rather intriguing that whenever there is a debate on reducing "defence expenditure" almost every expert talks of cutting on the standing army, per se. Why is it that the 'idle' work force of these 'loss making' Ordnance Factories, busily manufacturing obsolete, technologically antiquated equipment that nobody wants, has not caught the attention of our policy makers. Other than the "vote making" potential there does not seem to be any conceivable reason for this lack of concern where our vital national interests are concerned.

Another enigma is that successive Indian governments and the defence bureaucracy have fostered the idea that defence production has been reasonably successful over the years and that slow but steady progress is being made towards self-sufficiency. In fact that is not the case. In many instances the public sector undertakings have either not performed well or been frustrated in their attempt to do so by some vested interests. Many of the claims that systems are 'indigenous' are in fact misrepresentations, as increasingly the term is being used to cover production which involves little more than assembly, where the local content is minimal. Over the years there have been several instances of inefficiency, mismanagement, serious delays in the execution of projects, shortfall in production of vital defence stores and under utilisation of capacities. Despite the views of the Indian Government, several Indian pundits and a good many external analysts, the Indian defence industry would seem to be in a poor shape and clearly unable to reach the standards which its architects had envisaged. Indeed, India's ability to pass off equipment produced using copious amounts of foreign technology as homegrown is an extremely significant propaganda exercise. A classical example is that of MBT ARJUN where a critical report on the project in 1990 listed the following areas which relied upon foreign technology: (a) engine from MTU FRG; (b) transmission from RENK also FRG; © FWM fire control-FRG; (d) primary sight OLDELFT from the Netherlands; (e) tracks from DIEHL from FRG; (f) TCM hydropneumatic system from the USA. With all these imported inputs what exactly is indigenous becomes difficult to identify. With so many different countries and their 'defence industry' contributing such major sub systems one finds it virtually an impossible exercise to achieve self-sufficiency. There is any number of reasons (other than inefficiency of our PSUs) for this kind of situation. Our scientists are a party to this misplaced faith and confidence in our capabilities. As a symbol of our technological advancement and in our bid to reduce reliance on others we often task our R&D to come up with solutions to more intricate problems. Often they also give assurances and make promises they know that they cannot keep. Every body plays along in the fond hope of achieving self-reliance. There is another and more important reason for this situation. Much of the blame for the lack-luster performance, ... would rest with the armed forces. The explanation for this antipathy may be quite simple- Indian officers are much less enamoured of Indian defence technology than might be expected... the majority of Indians are rarely enamoured of indigenous products such as cars and electronics. *To compound the problem there are any number of instances where the user has been unwilling to repose faith in a local product. Foreign experts have generally identified the Air force as the main party that has always felt that a domestic product is not good enough*[1] (italics added). Unless, we ourselves develop faith in our equipment and insist on quality improvement if something is wanting it will be very difficult to find buyers from abroad. Secondly, we must evaluate our capability realistically in certain technological fields before tasking our scientists to come up with solutions to our problems.

What do We Do with our DIB?

An unhealthy defence industry is neither good for the military nor good for the nation. Inspite of the ' political and bureaucratic justifications' there is still a very wide gap between the actual and economical use of industrial and technological infrastructure built over the years at considerable expense and its functional and practical utility to military. To cap it all there will always be an urgency to keep the public-sector undertakings going for reasons other than efficiency and economical viability. The subject needs a detailed discussion in its own right in order to arrive at an answer to avoid the wasteful under utilisation of the installed capacity of these PSUs. To date, there has been no policy debate about efficiency, competitiveness, and the benefits and costs of our defence industrial base. That, however is not the main thrust of this work. Therefore, it would suffice to say that if the government is not already examining this aspect of economical utilisation of these units, it is time it did so. There is an immediate need for a total restructuring of our "defence industry". That is, if we want to make full use of the resources already created we must bring in professionalism and commercial instinct. Besides giving confidence to the customers such an approach can bring in an element of pride in the workmanship where the employees of these establishments are concerned. Once a decision is taken and implementation undertaken some kind of via media between the rugged defence needs and normal civil usage can be found so that these Public Sector Undertakings can also start showing some profit instead of the losses as is generally the performance today. The following measures could also help:-

- In order to meet the requirements in a stipulated timeframe and in order to equip the troops, in addition to the Ordnance Factory productions, which usually are behind schedule, huge quantities of stores are being bought from the private sector as well. Ordnance Factories also buy a lot of material and components from civilian industries, besides tapping them in areas of key technologies. By transferring some of the traditional work to the civilian sector a lot of duplication can be avoided. This will lead to the defence sector becoming smaller and specialised, requiring much less funding.
- There are certain things like arms and ammunitions that are better left to the public-sector for obvious reasons. This monopoly carries with it the liability of concentrating on the production of newer and more difficult and intricate arms and ammunition, and the Ordnance Factories should concentrate on this aspect. But there is no real need for the defence services to be making their own clothing items that are, in any case, far too expensive, coarse and ugly when compared with commercial products. Vehicles ex Vehicle Factory at Jabalpur are neither fuel efficient nor technologically

modern in design or performance. In addition, we have heavy production slippage leading to a situation where the Army has to buy a large number of vehicles ex-trade. Today, due to this induction of a vast and diverse range of vehicles the services have to hold a very large and diverse spare-parts inventory, besides training its fitters in handling that many types of vehicle. In times of war this could be a serious problem because one cannot really carry so many varieties and types of spares or expertise. If vehicles, particularly the load carrying, ambulances and certain other varieties that have civil-end use are brought on a common platform, the need for holding huge inventory of spares will decline. It will then be possible for the trade to meet the requirements in most of the places where the military is expected to operate.

- There are a large number of items in service, which have civil end usage and specifications. The military, however, continues to insist on specifications formulated years ago for no particular reason other than the tediousness of introducing fresh parameters and designs even though it is accepted that the later date products are more economical and user friendly. By designing our domestic implements in a manner which measures up to the defence needs as well, we will be saving a lot of money besides creating a large number of jobs involving different kinds of trades and technical skills. This will also improve the availability of such items, thereby enhancing the confidence of the troops in their material support complement.
- Only those defence establishments that make items purely for defence use can continue to do so. The areas of exclusive defence need should be quickly identified and appropriate establishments be given time and target bound orders. Where required, technologically should be updated, and if security considerations permit involve commercial sectors even

in those cases where sensitive equipment is being produced.

- Establishments like the clothing factories, vehicle factories and some of the others can surely diversify (as some of them like Vehicle Factory at Jabalpur are trying to do) and take on the civil needs in their range. A reason for not doing so could be that they lack marketing capability and are not able to produce large number of items at competitive prices. This may be so due to their preoccupation with meeting the rigorous technical requirements of the government customers, or the usual lethargy of a PSU. But these are not insurmountable constraints. At one time, as the old timers tell us, people had a lot of faith in a product from a government factory. By being competitive and by the induction of an element of professionalism these factories can regain customer confidence.

In USA where trade meets most of the defence needs a policy exists where the industry can sell its product in the world market once defence needs are met. We could also evolve such a policy giving adequate freedom of action to the industry to make their product economically viable. We have been shy of giving our civilian industry any kind of elbowroom even where we lack capability we try to do things on our own within the ambit of 'public sector'. Only after failure do we look elsewhere to cover technology mismatch. We cannot really set up a production line for every item of military need as certain more advanced countries tend to do, either individually or in collaboration with others, because they have well established markets. In our case it will never be cost-effective unless there is an export potential and/or civil end usage. In almost all the hardware exporting countries, for selling pure military hardware the government's sanctions are required. But in the case of products that basically have more of civil end usage the governments are not particularly rigid in their approach. This way US, erstwhile USSR, France, England, Sweden, South Africa, Japan etc.

have shown how the defence industry can support the economy as also earn revenue through exports of weapons and technology. We must understand that this approach on one hand provides support to the economy in terms of higher utilisation of production capacity, higher income and employment, and on the other it helps reduce the relative burden of the defence sector on the rest of the economy. Somehow we never accepted the fact that defence oriented industry contributes substantially to the national industrial production and foreign exchange earnings. We should formulate a deliberate policy to involve our civil industry, and also explore possibility of setting up joint ventures with friendly countries so as to make defence economically viable.

Private Sector and Defence

The task of equipping the armed forces is all encompassing. There is hardly any walk of life that does not influence the iogistic needs of the military or is not influenced because of this need. Therefore, one question that straightaway comes to mind is- is there any obvious connection between the national industry, and defence? Surely there is such a connection. There should be no doubt that a very real connection exists between the most effective defence and a vibrant industry. This connection must be looked at from the military point of view as well as the business point of view in the overall national interest. In a democracy a crucial aspect in the total defence effort is the role of the civilian sector in national defence. No military can exist in isolation.

Defence requirements are generally spasmodic and have a tendency to be unpredictable, in so far as a wide range of purchases is concerned. In more advanced countries the civil industry primarily meets the defence needs. It can, therefore suitably build its production cycles and schedules so as to incorporate both the military's and the civil customer's requirements. It is also possible for them to explore foreign markets for selling their products when there is a slack in either the military or civil demands. Hither-to-fore, in our country in matters of defence production private initiative was not encouraged on the grounds of secrecy, inadequate (?) technical expertise, bureaucratic role and control; and the almost overshadowing presence of the Public Sector Undertakings and Ordnance Factories. Not surprisingly, we have been rather slow in realising that a crucial aspect of the total defence effort is the role of the civilian sector in national defence. Only recently have experts started to point out the contributions that can be made by the civilian sector to defence. Involvement of private sector in defence has immediate effect where emergence of new technologies such as electronics, avionics, computers etc. are concerned. These technologies enable existing industries to develop newer and wider range of equipment, instruments and materials. Computing machines, control devices and synthetic chemicals are examples of private investments' being so induced to create new facilities or modify old factories and production equipment. In turn, the tools and materials created by the new technologies make possible economies of production in other industries, calling forth-new investment to finance cost-saving innovations and increased output.

Historically, in India there have been few incentives to integrate civilian and military engineering and production. It is only in the nineties that emphasis has been placed on exploiting the complimentarities in the defence sector and the rest of the economy as also to augment export earnings of our defence and military products. Because we did not integrate our defence needs with the country's industrial, scientific, and technological potential we have been unable to take full advantage of the industrial and technological evolution that has been taking place in more advanced countries of the world. This has created a piquant situation. In that, for most of our military requirements we traditionally look for imports because neither the civil industry nor the defence industry has been geared for mass production of a wide range of systems and subsystems which could be used both by the military and the civil. In the absence of a heavy buyer like the Military, industry has generally not invested a great deal of money or expertise into modernisation of its production facilities .Resultantly, we have not been able to exploit the growth potential of our economy, technology and industry in a manner that could ensure a better defence preparedness with very little strain on the country's resources. This has adversely affected the industrial capacity, production lead time and price of the product. This has also resulted in a situation where for technological or tactical upgrades we have to look up to foreign suppliers; be it the field of electronics, avionics or armament.

With the passage of time civil industry has built up enough expertise in a very wide field of technology, it has gained enough confidence to make meaningful contribution towards defence. An integrated approach of indigenisation, reverse engineering, utilisation of available and absorbable technology in the civil sector offers the best prospects for enhancing military and industrial capabilities simultaneously. This means orienting and mobilising the industry in such a manner that self-reliance is achieved in as many fields and as early as possible. To boost up the process the government could even underwrite a certain portion of the expenditure, or become partners with reputed industrial houses for research and development in areas which are common to both the military and civil. Once a deliberate and collective decision is taken in this direction it will become obvious to the decision-makers that in the long run the route of inter dependence in respect of major programmes will become increasingly attractive to the trade as well. However, such "pump priming" may well have some very important long-term effects. The military's emphasis on high-performance devices can create tension between its needs and the commercial interests of the supplier. An unfortunate consequence of this situation is that some kind of built-in opposition to the development of indigenous defence oriented industry in the civil sector has crept in. Another problem that we have is an uncertain demand pattern, regulatory controls, and lack of freedom to maneouvre and stringent security requirements in some cases that inhibit the private sector participation. They, in any case, feel that due to less profit margins, red tape and inconsistency of the user there is hardly any motivation. This can be rectified by allowing industry to suitably alter the product for the civilian market. There is no doubt that there is a need to at once formulate a specific policy for involving the civil sector in our defence needs. Another thing is that we are reluctant to pool the production resources available within the country both in the public and private sectors so that necessary capacity can be created. Because of such a mind-set we try to establish 'self-contained' units of production. These often get bogged down because of a very heavy outlay involved.

Because of our dogmatic approach and mutual lack of faith for unknown reasons the firms working exclusively in the commercial sector believe (and not wrongly), that doing business with defence is just too difficult. Even in 1962 when there was an urgency to equip our forces in hurry and the government was exhorting the industrialists to come forward and help found very few takers for the "national cause". A few industrialists who, not withstanding the handicaps came forward to assist were frustrated by the procedures for placing orders by the Government. This is not particularly conducive to the private sector exhibiting an active interest in providing assistance to the military in equipping the forces. The basic problem is that a manufacturer who invests money and effort to provide items to the military expect commensurate returns on his outlays and the risk involved in shifting his focus from a known civil market. There may be valid arguments for the present situation, but these have to be addressed early and we should be developing new procedures to encourage integration of the civil sector with the defence if we want to have an affordable security along with development. To overcome the inhibitions in both the sides joint exhibitions where the industry bigwigs actively participate have recently been held. Industry has identified and highlighted specific problem areas. Mainly, their grouse is that the private sector is not allowed to participate in the armament industry even though the field is wide open to foreign companies. The Army expects prototypes to even surpass the rated performance and to be perfect. The industry finds it difficult to participate because the Army is known to make mid course changes in its Q.R, administrative lead-time to finalise an order is excessive and bears no relation to the production time given to the industry. thereby leading to delays and even obsolescence of technology. The problem is genuine but not insurmountable. It is important to understand that the logistics needs of military often have a long gestation period. And things cannot always be bought off the shelf. It takes considerable time, energy and effort to identify a suitable gun; an appropriate aircraft and ships that meet the country's defence requirements. Identifying the appropriate industrial sector that can provide spares and subsystems in due course is even more difficult. The process of absorbing and assimilating any, particularly a new technology, be it military or civil is considered fundamental to development. This is a time and effort consuming exercise. Planning and execution of orders are usually spread over a number of years; in the meanwhile, because of the pace of technological progress, it could happen that a rival equipment claiming to be superior may come in the market. We must be prepared for such a situation when one has to make a careful judgement so as to decide at what point of time the ongoing procurement will become obsolete. Before altering an earlier decision a thorough examination of the relative merits of the old and of the new for the tasks the equipment is meant for be undertaken. In case it becomes essential to alter an earlier decision then the manufacturer who had been assigned the order earlier should be given first option to offer updated product. If we want active participation of industry we have to gain their confidence as well. For achieving this end the right way will be that other than the computer industry (where newer technologies are emerging every day) we should seal the agreed upon specifications at the time of placing our order on the industry.

From the military perspective it is argued that its "unique" requirements and its specialised procurement and accounting procedures are dictated by the fact that public funds are being utilised and they, therefore have to be extremely careful. This variance in perception leads to a situation where the manufacturer's view point and that of the Government's representative (invariably a bureaucrat because all financial powers rest with them) who seemingly is anxious to safeguard the public money, rarely agree. The situation could be improved by giving appropriate financial powers to those who are answerable to the nation and to the troops. Because the bureaucracy does not want to let go the stranglehold it has on the military no satisfactory solution for this situation has yet been evolved. It appears that the void between the Trade and the Military is too wide and old to be filled in a hurry. We have to break the current thinking regarding involvement of trade with defence and give confidence to the suppliers about the seriousness of the government's intentions. A confidence building exercise will have to be undertaken where already identifiable lacunae will have to be removed in the first instance. The military has to do some serious thinking if we want to keep our defence expenditure within reasonable limits and defence has to become a front runner in bringing in development in the country. For achieving this aim we could also take the following measures, as was done by the US government in the late fifties in order to boost defence oriented industry:

- **Guaranteed markets at guaranteed prices.** Commitments were made for Government purchase of output when a project involved an abnormal risk or when an assured market was required to induce either investment or production. In some cases, the commitments consisted of contracts to buy the output of a particular project, if it was not otherwise sold at a negotiated price. Sometimes even premium was paid to the potential suppliers in order to motivate them to meet military needs.

- **Direct loans and guarantees of commercial loans by the Government**. If private financing was unavailable for expansion projects because of the business risks involved, the Government made direct loans for investment in facilities or working capital or guaranteed the loans through banking channels.
- **Financing of plant construction and equipment through military contracts**. The initial costs of equipping, and in some cases, of erecting facilities to produce items of defence use under military contracts was supplied by the government. These amounts were covered in the contract price for specialised military equipment and in due course of time with foreign sales and some times with the civil end usage being found the facility got strengthened. The government recovered its money in the form of the product and with assurance of firm supply if and when required.
- **Installation of Government owned equipment and construction of Government owned facilities**. Production equipment owned by the Government can be installed in the contractor's plants when the contractor is either unwilling or unable to bear the cost. In our case we could do the reverse - by inviting the industry to set up plants in those PSUs that have become redundant due to antiquity of their machinery and non-productive work force.
- **Grant for research, exploration and development**. Grants can be made to stimulate development of new methods, which can lead to greater outputs or conservation of scarce materials.[2]

A combination of the above along with any peculiar requirement of a specific industry keeping the overall national priorities and interests can pay rich dividends in so far as military needs of material are concerned. We should formulate an industrial strategy and make a specific defence-oriented programme at the national level so that the dual goal of national prosperity and adequate defence can be achieved. Military, on its part, must make full use of international technology and markets to create and nurse a civil industrial base within the country. This will cut on costs that will then be dispersed over a larger base while also fostering standardisation

Having identified the problem areas that restrict industry's active participation in meeting defence demands and in order to overcome these we should set up a corporate body to coordinate the whole gamut of identifying defence requirements and those segments of the industry who have the capability to measure up. Closer interaction between defence services, PSUs, Ordnance factories and civil industry in terms of conduct of user trials, professional promotion of sales and sharing the human resource for mutual benefit be evolved. Revamping of the "public sector" and making them more competitive where pricing, quality of the products and user confidence are concerned will also help. The structure of this organisation should be evolved after due consideration. The manufacturers in private sector (who are to contribute towards defence equipment), defence sector including Ordnance factories and other PSUs and DRDO need to be brought under a common umbrella. This apex body can also facilitate export because in so far as the civil industry is concerned its participation can only be profitable and meaningful if the industry also undertakes exports. Where exports are concerned an aggressive marketing strategy for defence exports needs to be evolved with active participation of our foreign missions, increased exposure of our industry and establishments to foreign missions in our country, and participation and organisation of exhibitions.

Defence Exports & Industrial Development

Many nations use their defence exports to augment their sphere of influence. Exports help their defence sector pay for itself as no country can afford to have appropriate defence unless its products also have an export potential. That is why they invest heavily in defence research and production, the other reason being a quest for gaining strategic advantages over potential adversaries. Even China has been reaping substantial benefits through arms exports. In our case, due to our foreign policy posture export of arms has been considered to be unethical and no conscious effort was made till 1991 to really convert this into a source of foreign exchange earning or as an influence-augmenting option. It however seems unlikely that India will succeed in becoming a major exporter of defence equipment, for several reasons. First, the market is saturated. The international arms market is in a considerable slump, which ironically dates from the time when the Indian Government decided upon the shift in policy. Particularly hard hit have been sales of major weapon systems, which is precisely what India wants and needs to sell to make the necessary gains in foreign exchange and to improve its standing amongst developing countries, especially in the neighbour-hood. But the market is now much more geared towards technology, which India would find difficult to supply because we haven't encouraged the civil sector to invest and the government has always been strapped for cash in so far as defence related technology is concerned. Second, India lacks the expertise, and the experience of its competitors and the necessary staying power to effectively compete with the established players in the field. Also, the government does not have competence as entrepreneurs.

Secondly, (by definition) in India both the sales pitch and the decision-making process are so slow and bureaucratised that they lack the flair and innovation of competitors from the other arm-selling countries, or for that matter the vigour available even in the private sector to push its product is missing. For the market forces to operate more effectively in the defence sector, many of these barriers to entry and exit, which in no way contribute to organisational efficiency, must be greatly reduced, or better still, be eliminated. Another, and perhaps most important reason is that the equipment that we make, by global standards is not of the highest quality. There is a need for a very deliberate effort on the part of defence industry to, firstly satisfy the Indian Defence forces with regards to their products. Unless, the Indian Army itself buys domestic products there is no way the industry can convince a foreign buyer with regard to the quality of its equipment. Once they have satisfied own troops they can go in for aggressive marketing, preferably outside the ambit of bureaucratic controls and interference. By involving the civil sector in this endeavour we can bring in a faster rate of customer confidence which so far has been lacking. To enter the era of self-reliance where both the public and private sectors participate and the imaginary barrier between "defence and development "is removed we could do the following:

- **Develop defence oriented civil industry**. By its very nature and competition the civil industry is more contemporary in order to survive and thrive. Identify those sectors of civil industry that have potential for assisting defence. Defence should come forward and do this identification in consultation with the biggest players in the field. To absorb defence needs the industry may require importing certain machinery and equipment. They may also have to send some of their professionals' abroad for training. Whenever the industry has to import machinery for indigenous manufacture they be assisted in obtaining all the requisite clearances from DGTD in a time bound manner and framework. In all probability the product will have 'civil end' usage. Often enough it has been seen that in such a situation the concerned bureaucrats tend to delay matters because in their opinion if a vendor is to make any profit then he should follow the classical 'red tape' route. This in itself is reason enough to put off most industrial houses. The 'Planning Commission' or another high level body, that includes users as well, could handle such tricky situations. This will also give confidence to the industry Vis a Vis the government's intentions and may hasten the process where the industry would like to contribute to the defence effort.

- **Indigenous/imported raw material is made available on priority basis**. Bottlenecks in the production of raw materials and sub-contractor components lengthen the production time, besides leading to wastage of time and energy. The organisation should ensure that this does not happen. And where the raw material is required to be imported the defence should assist them in doing so. And if, the end product has a civil end usage in addition then the military user should form part of the committee dealing with the subject even if the end product has been committed for export. However, this will need a lot of pragmatism, both on the part of the bureaucracy and the industry, because cussedness happens to be a favourite pastime of the mandarins. (A long time back at a 'Corps Headquarters', problem with regard to special tools for an imported weapon-system had become serious. Import required foreign exchange and the replacement of tools being rather low in the order of priorities in those hard days of scarcity; there really was no chance for the tools to be finding any place in any list (of priority) for purchase. The situation (to put it mildly) was critical and every commander was concerned with the requirement of finding a speedy solution to the problem. It was found that these tools were actually manufactured by an Indian firm located in Faridabad. The firm was exporting them and we were buying them against dollars from the country supplying us the main equipment. Our efforts to procure the tools from the firm failed because they were allowed to import steel of a specific grade to manufacture goods for export only and that is what they were doing. And if the army wanted their product for domestic use then following the proper and the prescribed procedures was the way to go about it. In that, the defence ministry should approach the steel ministry, who in turn should obtain clearance from the finance ministry for allowing the firm to violate the operative clause which explicitly stipulated that the product made from the imported steel be exported only and not sold within the country.) There are any number of instances of these kind of wasteful and avoidable situations in almost all arenas of our material support to the military. An appropriate agency should deal with such impediments, which in any case are not in the national interest. The 'coordinating agency' recommended earlier and at the national level can effectively break the control of rigid rule bound controllers that we have to live with.

- **Associate trade with the reverse engineering process.** For want of adequate 'in house' expertise within the military it has often happened that we have either not been able to unravel a particular technological product or have taken too long in doing so. It is not the fault of resident engineers and the technicians. Technology is evolving rapidly. The vast majority of today's engineers in government service having completed their education a long time ago are clearly outdated and professionally obsolete. So that technical expertise available in the country is jointly and appropriately utilised we should associate the concerned branch of industry from the very beginning. By involving the trade where, by virtue of the fact that people have to be current in their technical awareness for their survival and growth, we could speed up the process. Also, by allowing the civil industrial sector to commercially exploit the latest technology we will be offering enough incentive for speedy development of the item or equipment. There are a large number of products, particularly in the field of avionics and electronics, which have both military and civil usage and industrial houses could find the effort worth their while.

- **Industry is given incentives and encouragement to go in for defence related items.** Tenders need to be made attractive. In addition, the suppliers should be given assurance with regard to definite quantities to be supplied. The main thing to be guarded is that after a particular bidder has invested so much money and effort in developing a specific item for defence; he is not deprived of his legitimate dues. It often happens that because of policy changes and other reasons some smooth operators come up with cheaper versions. Their profit margins could be low because they hadn't invested in R&D and could therefore afford to quote lower than the pioneer. Fully knowing that innovative technology always costs more than imitative technology we, due to our archaic rules and procedures, let it happen day in and day out. The " lowest quotation syndrome" has been doing more damage than good not only in the defence sector but also in various other walks of life. These speculative intruders are often encouraged by the 'financial representatives' in the fond hope of saving some money for the state. There, however, is no guarantee that the new competitor will be in any position to deliver the goods as per the specifications and on time. There is a penalty clause in all the contracts. But it is hardly ever invoked. In any case, whatever may be the action taken in pursuance of the penalty clause, the stores do not materialise and the troops suffer in the bargain. If we want to discourage the interlopers the pioneer manufacturer should be given assured orders for quantities that are economically viable. These supply orders must be of such a magnitude that these are sufficient for him to cover the costs of development and establishment of production line. The aim should be that the pioneer sees a reasonable profit on his investment.

- **Refine procurement procedures**. Industrialists who come forward to assist the military are often frustrated by the procedures for placing orders by the government. Their general problem is that the Government's representatives, particularly from the finance side rarely ever agree with the user's perception or that of the supplier. Today our procurement procedures are so complicated and time consuming that a large number of potential suppliers shy away from coming forward. They do not have the time and patience to wait in the corridors to have an over busy bureaucrat go over their bids. Then we have so many recommending and approving channels that the initial bid often gets time barred or the prices go up because of policy changes and the supplier is no longer in a position to honour the offered price. What is more intrigûing is that even commercially proven products while being considered for introduction in the Military are subjected to so many trials and tests, that many sensible people prefer to stay away from bidding where a military requirement is concerned.
- **Public and private sector could invest in R&D**. Accepting that the concept of "induced innovation" results in R&D objectives having a distinct influence on the evolution of technology, joint effort, will lead to improvement of products which, while enhancing the military preparedness will also help the industry in finding better consumer acceptance. Integration of defence and developmental R&D will lead to economies of scale in regard to production of high-tech materials and equipment. Dual use for the research programme (vs. Defence only) is likely to influence the decision as to which technology gets emphasised and how the technology evolves. Concurrent exploration of usage in both the military and civil fields will help in

achieving speedy results, because industry could smell commercial profits and growth of its strength. This, in itself, will stimulate the industry to come up with end products much faster. This may also lead to obtaining some returns on defence out lays, besides providing higher involvement of those who still have productive years left- such as utilisation of services of ex-servicemen, area development around border roads etc. We could integrate naval and civil maritime activities, civil and military aircraft production, etc. and thus save on overheads in a big way.

- **Develop an 'integrated' R&D institutional network at the national level**, in such sectors as electronics and telecommunications where innovation is increasingly driven by business applications. This will optimise the expertise we have in both the civil and defence sectors, as any R&D in isolation is going to be prohibitively expensive and, more often than not a duplication of effort. If such an agency is set up under the aegis of the 'Planning Commission' the prevalent problems of lack of interaction and coordination besides duplication, can be overcome. This can be particularly useful in the case of research related with life saving drugs; signal communications network, and satellite communications, computer based solutions etc, which have both civil and military usage: (In USA the American government played an important role in supporting the crossover of technology from the military sector to the civilian sector in the case of transistors. By sponsoring research, subsidising engineering development, funding the planning of production, being the initial "guaranteed", buyer, standardising practices, and disseminating the results of such work, the Army Signal Corps played a central role in defining the structure of the early transistor industry.)[3] Sufficient design and engineering teams must remain embedded in the structure to produce new components and systems on demand. They must keep innovating and applying technology as it evolves. That means our experts from the combined R&D will have to keep an ear to the ground and be alert to incoming technologies well before these become commercial properties elsewhere in the world. Having such an 'in house' expertise, besides giving greater confidence to the users creates greater awareness among potential purchasers of our equipment. Finding buyers other than defence will create market for our products bringing in revenue to the state besides strengthening our industrial base in a relatively economical manner.

Industrial Mobilisation

The economy can be made stronger by improving industrial capability and capacity, by taking those policy decisions that help growth, and by using means where the citizens pay their taxes and strengthen the country's economy allowing the government to pursue pre-designated role towards achieving the goals. This process again depends both on the available, and anticipated production capacities and technological developments of industry, and on having people prepared and willing to make sacrifices in the national interest. The economic act of industrial mobilisation is related to the grand strategy. If industry cannot provide the necessary equipment (and there is no adequate and reliable external source), nor is there R&D capability to undertake design and development, then the strategy must either be abandoned or fail. In today's uncertain strategic environment there is a high probability of facing a "no plan" crisis situation in so far as material support is concerned. Even if a contingency plan were to exist, that the assumptions and resource estimates would match is an uncertain factor. Actual mission, enemy, terrain, troops and the time available may not allow this. Unless we gear our industry, both in the public sector and the civil urgently and provide suitable guidelines, rapid industrial mobilisation cannot be accomplished from a standing start without prior planning and without a baseline defence and defence oriented industry. In order to provide a total and economical logistical support, two critical aspects have to be taken into account- the industrial build up and the total mobilisation of the national economy. Central leadership and coordination are needed for this purpose as much as in the energy and environment areas. This also means that the task has to be assigned to a central body who should periodically monitor, and where required, give assistance and confidence to the industry. A 'long term' government policy and suitable agenda where the military and not the "defence bureaucracy" has a say needs to be prepared and published so that all concerned know what to expect and what is expected of them. Meaning, we have to develop a national action plan to achieve mobilisation readiness and pursue that plan where we create an "mobilisation base".[4]

There are certain key factors which have a direct bearing on industrial mobilisation. The first is 'time'. The time it takes the Nation to get to mobilisation production rates must be offset by war reserve stockpiles. The less time required, the smaller the necessary war reserve stockpile can be. The second factor is the inevitability of obsolescence of military equipment. If we build too big a reserve, then besides locking up capital we also run the risk of not finding adequate spares when a need arises. In case industry is associated with defence preparedness then it can produce even a current item at a much faster pace. On the national level we must recognise how difficult it is for the operating forces and the industrial plants of the nation to acquire momentum at the start. It takes great deal of time and effort to build up the advance area stockpiles necessary to support sustained offensives. It takes even more time and effort to achieve the industrial momentum that is the foundation of this logistics builds up. The process of changing the rate of production is often so long that it may not be possible for the supply system to respond quickly to the change in demand even when the urgency is recognised. Another inherent danger is that the supply system may be geared too closely to peace time operation and it may not be either technically or organisationally prepared for the very great changes that war brings. The problem is intimately related to the study of usage data planning factors, the logistics snow ball, to flexibility and to readiness.

For a mobilisation plan to succeed The infrastructure of the country must be maintained and upgraded in areas such as communication, road & rail system to provide a world class distribution network. In the absence of a reliable and efficient system it will always be a problem to react to an emergency. As has been said earlier we should identify potential bottlenecks in our communications system and take corrective and rectification measures. For example, where transshipment of military stores from broad gauge to meter gauge is involved the gauge conversion should be undertaken as early as possible. Otherwise hundreds of tons of military stores including ammunition will keep on lying at way side stations awaiting "railway's convenience" for move forward of these stores. In the interim the troops will remain uncertain of their logistics requirements materialising at the desired place and time. Similarly, where heavy road traffic is envisaged due to movement of military stores and heavy equipment the road network be improved, where required 'by-passes' be created to avoid problems of traffic controls. Even road bridges etc en-route should be also strengthened.

How do we Cut Inventory Costs?

Army alone holds an inventory costing more than 50,000 crore rupees consisting of over 5 million items. Recurring costs including inventory carrying costs eat away a fairly large chunk of our budget. With a much improved infrastructure and technology base we can make heavy reductions in our holdings and provide for mobilisation of our industrial base instead. If we keep in focus that a healthy production equipment industry is a key factor in the nation's ability to maintain a healthy economy and an adequate mobilisation base, it can be done. We can identify the relevant industry sectors now when there is no specific pressure associated with a war-like situation and assign them the task of supporting military needs during normal times and also give an indication of our likely operational requirements. The industry and the military will have to decide on a mutually acceptable sequence for establishing a time frame where the 'lead time', warning period and even likely duration and location to make the stores available can be decided upon. A committee can be formed under the "Minister of Defence Production" to establish criteria for the planning programme in order to achieve the following:

- **Providing for Efficient means of Procurement**. Efficiency and therefore economy in production should form a basic parameter for assigning the responsibility to a potential supplier. (Why the products of government factories and PSUs are costlier has been discussed earlier). More often than not, once a defence order is placed the manufacturer sits pretty. (This is more common amongst the PSUs and those welfare oriented government approved concerns who are not necessarily concerned with the efficiency of military) because he is sure that this customer is not now going to look elsewhere for his requirements. Notwithstanding the problems the arrangement causes to the troops who are dependent on the service organisation for their needs, he could always obtain a "stay order" from a court and restrict the military from finding a more pragmatic source to meet its urgent requirements. A systematic and critical analysis of the production house before an order is placed will be a big help. A careful selection of the vendor by critical inspection of his premises, his financial, technical and managerial capabilities will prevent disputes from arising.

- **Providing or increasing the capacity to produce these selected items**. An accurate forecast of actual requirements spread over a specified time frame is to be made. The 'provision figures' thus arrived at will tell us whether there is a need for setting up additional facilities or that the targets can be met by increasing the capacity in a given time span. The factor of likely obsolescence will have to be kept in mind. If the item(s) has civil end usage this problem can be easily resolved. Extensive use of material from the commercial market will help civilian economy in a big way because the supplier knowing of the big spender that our military is will be in a position to create additional distribution and production facilities. Rates will get to be more competitive; quality of the products will also improve. Some problems, however, are envisaged. In respect of civil end use items, the possibility of stocks disappearing from the market cannot be ruled out, particularly when the industry smells a big order for spares and sub systems coming up due to some operational urgency. It would also be difficult to tap a large number of dispersed small traders and stockists where spare parts are concerned. Secondly, and more importantly, for some specialised and often high priced items there could frequently be only one source of procurement. This "sole source" could become a hindrance to rapidly increasing the availability of the items when the demand suddenly goes up for some reason. The production facility manufacturing a particular item, in some cases, cannot increase its production fast enough to meet the rapidly rising military requirements which traditionally has cyclic surges. Some manufacturers may not deem it feasible to expand their production facilities, at great expense, to meet temporarily increased supply to the government. In order to avoid such a situation, it will be appropriate for the military to identify those items that could become problematic at some stage in the equipment 'life cycle' and to develop more than one source. If civil end usage could also be ensured by standardisation then chances of more suppliers coming up also increase

Assessing the tradeoffs between war reserves and production base. This will be a very deliberate effort on the part of all concerned. All aspects of holding reserves and therefore locking up capital for indeterminate periods will have to be weighed against having a production base that can be activated in a prescribed time frame. The military requires its own base structure to maintain a surge capacity to support limited conflicts. But that capacity should be small and geared towards essential material such as ammunition, consumables and repair parts not in common usage. Other than these items it should be possible to procure essential requirements ex trade. Over 50 percent of our inventory, be it in the field of electronics, engineer plants; vehicles or clothing has civil end usage. Military could identify and interact with the reputed suppliers in the border towns so as to ensure supply locally till an operational situation stabilises. By shedding provision of some of these spare parts and subsystems to the industry in peacetime we can make a deliberate beginning towards a thinner inventory holdings at various levels, particularly 'field formation's . This action, in itself, will provide an incentive to the industry to undertake production of defence-related items. The biggest problem will be that of ammunition and explosives. These have limited shelf life ranging from five years in the case of some 'primers' and 'fuzes' to as much as thirty years as is the case with the 'artillery shells'. The items have virtually no civil end usage and the production base cannot find market unless the products are exported. Holding huge reserves will be economically impractical.

- **Maintaining an adequate production base**. A production base gives confidence to the users and provides a medium to the planners to arrive at more realistic base to work out their 'provision' figures in a manner that besides ensuring economy also provides flexibility in so far as assigning priorities are concerned. It should have surge capacity...

Indigenisation

Procurement of military hardware from a foreign country is in a large measure governed by the mutual political relationship between the two countries. This means that in the formulation of its defence plans a country has to manipulate its foreign policy so as to keep these avenues of supply open. During the "cold war era" and due to certain political compulsions we have been heavily and largely dependent on the 'East European' countries in so far as our defence needs are concerned. We did not realise that if a country cannot produce for itself the systems and infrastructure considered essential for modern defence, it will be forced to look beyond its national industrial base for the relevant technology and expertise. We also did not appreciate that a country depending largely on imports of defence equipment virtually subjects her defence and foreign policies to the veto of the suppliers who in the interim, could have had a change of heart or get fragmented like the USSR. This dependence on others has been a major reason for our own defence industry to be somewhat slow in bringing in self-reliance. We did not invest enough in the industrial infrastructure which could provide military with indigenous spares and accessories where imported equipment is concerned. Nor did we encourage the civil industrial sector to assist. Earlier we did not have a very well developed industrial base of our own and had to seek equipment wherever we could get it from. The changed world situation demands that we make deliberate effort to achieve self-reliance in as many military technologies as we can, and as quickly as we can because we now have a difficult situation where maintenance of a large complement of military hardware is concerned. This fragmentation of the USSR and shifted priorities of the suppliers where some of them either do not want to do business with us or want to revise terms and conditions agreed upon earlier has been creating tremendous problems for us in our endeavour to maintain the equipment imported from the Eastern bloc. The suppliers who were earlier following the dictates of Moscow no longer have any compulsion to do so. The foreign suppliers are apt to change their stand for a very wide range of considerations. In the instant case with all the suppliers having gone out of Russian control we have all kinds of problems where maintenance and sustenance of equipment already in service is concerned. A changing world scene since 1989 should have alerted us about forthcoming problems and remedial measures undertaken. Ostensibly, nothing of the kind happened. We need to appreciate that where foreign sources are concerned their level of reliability will always be low. Even in the case of free world countries where private industry handles defence requirements even the private suppliers have to toe their government's line in their own interest.

Having accepted the problems associated with depending on foreign suppliers we have been making consistent and persistent efforts in achieving some kind of self sufficiency. Generally we haven't achieved much success except for replacing a split pin here or an insignificant component there. To be fair, there have been some success stories, but the quest for self-sufficiency is far from fulfilled. An important facet is the fact that there have been too many attempts to indigenise in the defence sector when it is perfectly clear that the expertise and technology are unavailable at the national level. We tend to forget that technological feasibility and military desirability are not always synonymous. Further more, Indian decision makers have always tended to pitch requirements far higher than what the industry is capable of delivering- this is a complaint heard frequently from foreign consultants as well. This has, often, led to situations where the upkeep of equipment, its maintainability and sustainability has generally left too much to be desired. Delay in preparation of Initial Scaling Guide (ISG) of the newly introduced equipment due to non availability of supporting documents, like Illustrated Spares Parts List (ISPL) and the Manufacturers List of Recommended Spares (MLRS) causes slippage. Even if these documents become available, there are frequent changes in dependencies. There are changes in the system of provisioning, an irregular pattern of wastage, changes in periodic replacement policy, changes in the source of supply, and modifications to the basic equipment leading to a greater maintenance need etc., that cause imbalance in the policy for providing spares backing. They also cause escalation in prices and create uncertainties. The decisions with regard to indigenisation consequently suffer by default.

Selection of items for indigenisation is not based on any proper criteria and item indigenised is not as per the priorities laid down. Technical Committees (TCs) formed for the purpose select only 150 or so items of their capacity per year for indigenisation. Even then the projected requirement of the quantities is so little that neither the PSUs nor the trade finds it 'cost effective'. Reputed firms do not come forward due to meager quantities, fluctuation in requirements and procedural delays. Therefore, we should work out the liabilities for a period of seven years. Of which three years requirement is imported and the balance four years are projected to trade. To get over the problem of adequate quantity, which will be attractive to the trade, these committees should separately indicate items selected for indigenisation in the ensuing year and items that are amenable to the industry. Both these categories are to be cleared for only three years import so that a sense of urgency is reflected to all concerned and future requirements become economically attractive because imported equipment will normally have a life span of 25 to 30 years. In no case should any order placed on trade be cancelled. In order to make a tender attractive we could even order a quantity more than actually required. In the case of communications equipment and certain other categories like "snow items" etc the industry could be encouraged to identify even civil end usage. This will speed up the process.

Delay in inspection and trials of the pilot samples due to inadequate facilities with the inspecting agencies is another problem area. We have set up a huge 'Directorate General of Quality Assurance' (DGQA) which has its offices in various parts of the country, usually dealing with the specific items manufactured either by trade or PSUs located in that region. These agencies are normally not well versed with the technology of equipment which is 'newly introduced'. Often they do not have appropriate facilities either. For that matter, which set up is to carry out the inspection of a particular item or equipment is also not always identified well in advance. This creates an environment of uncertainty and causes avoidable delays. The absence of an integrated procedure leads to many activities being sequential rather than simultaneous. In the absence of parallel activities it is not uncommon for the supplier to be given a run all over the country, thus leading to an avoidable wastage of time and money, both at a premium with a private entrepreneur. There is a need for interface between various organisations involved in the process of indigenisation and active monitoring by a controlling agency.

Another reason for our failure in this direction is the fact that our Ordnance Factories tend to go in for misplaced economy when they undertake the task of indigenisation some important item. A classic case is that of manufacture ammunition for the 'Bofors gun'. The manufacturers supplied palletised ammunition where each pallet weighs approximately 600kgs. When indegenisation was undertaken, the concerned ammunition factory modified the original and the manufacturer recommended packing by substituting it with 'low grade' wooden pallets and instead of using 'rubberised' cushions used wooden stoppers. Consequently, the pallets broke in the journey from the factory to the 'central depot' itself. The factory sent its repair teams to the depot, but there is no way these pallets could withstand another journey to the forward echelon. That, they were economising on the cost of packaging without any consideration for consequences of their actions cannot in any way give confidence in the product to the troops.

Due to the reasons cited above, by the time the initial indigenous spares become available, the equipment normally passes its half- service life. Thus the benefits are not fully reaped. We need to integrate the indigenisation process with the life cycle of imported equipmrnt. To speed up the process a certain percentage of vehicles, guns etc. which have been extensively used for trials and in the training establishments should be withdrawn for experimental overhaul in the Army base workshops or at 'identified facilities' in the private sector. This will help in two ways, we will have more realistic figures for spare part planning, and the overhauling agency will be able to finalise the overhaul scales and gear up its resources prior to the start of the overhaul programmes.

In view of the wide variety of stores and equipment the armed forces need and the pace of technological progress that has accelerated the rate of obsolescence of equipment in the hands of troops, there will have to be sustained and well-planned effort for indigenisation over a long period of time if we want to achieve self reliance in the field of defence equipment. In order to overcome numerous problems in the realm of indigenisation it is recommended that before buying the equipment we should set up a 'project group' consisting of Ordnance, EME, Department of Defence Supplies, Base work shop responsible for the family of a particular equipment, representative of the financing agency, and the end user. The group should also include a member of the specific industrial sector earlier associated with same type of equipment. This group should carry out the selection of items for 'lifetime buy', identify those items which are likely to have a long lead-time if taken up for indegenisation. Only those items should be selected initially, for which the capability for indigenisation exists and not a particular critical item that needs to be indegenised so that a spare part is made available locally. Own capability to indigenise in an acceptable time span should be the guide. In order to achieve self reliance even when our basic equipment is imported we must make an all out effort to indigenise as much of spares as required, so that at no time will the equipment be wanting in exploitation for want of spares.

The Benefits and Costs of a DIB (Defence Industrial Base)

Various benefits of D.I.B have been suggested, some of which might differ between advanced and developing countries. Often these benefits are not explained carefully, nor are they evaluated critically. One facet of the 2nd World War, perhaps the only positive one, has been the rapid growth of technology not necessarily related to arms. There were so many things, goods and services just waiting for their turn to be discovered or invented in the scientific firmament, in some cases for a long time. Wartime need hastened their growth. Only one example, that of rocket-technology would suffice to illustrate this point. Perhaps the most important of indirect gains, however, are the spill over benefits from military research and development. Some of these benefits are obvious. The development of a good transport plane for the military can lead directly to improved civil aviation. New ideas resulting from research are likely to have numerous and unforeseen applications. In fact, it was military research that lead to such items as silicon transistors, improved anti-motion sickness drugs and new flame-proofing technique for fabrics. The latest revolution in information technology and worldwide communications via 'Internet' is originally a Department of Defence project in America. One can even say that wartime efforts eventually brought in satellite communications. Most of the commercially useful post-World War management techniques were developed by and for the defence sector, in an effort to ensure that the taxpayers would get the maximum amount of security for the money spent. Cost-benefit analysis, operational research, critical path analysis, programme management, programme evaluation review techniques (PERT), planning-programming-budgeting systems and system analysis are all examples of management tools that were developed for defence usage. Today, the industry is reaping rich profits because of these developments. Thus, the spill over benefits from research and development by themselves make it economical to increase defence expenditure where common man will also benefit in the long run. Our planners, politicians, bureaucrats and economists need to ask who gains and who loses from maintaining a national DIB; broadly, the benefits from a DIB include the following:

(a) **National independence, security of supply,(self sufficiency) and responsiveness in emergency and war.** This is often presented as a major benefit from a national DIB. It frees a nation from dependence on potentially unreliable foreign suppliers of essential defence equipment, particularly during a crisis or conflict. It also enables a nation to modify its equipment quickly to suit a particular requirement. Own arms production capacity also provides a form of status that enhances international influence. However, whilst governments might value independence, status and the self-sufficiency offered by a DIB, such an insurance policy can be costly if the DIB is to be funded exclusively by the state, or if the product does not have a market outside own boundaries. This means that there is an urgent need for more dynamic management of the DIB. We must remember that management innovation is an area in which defence expenditure can have a clear impact on national productivity.

(b) **Advancement in Technology**. The need to maintain a capability that a nation believes will be required in the future and meet its present requirements in a cost-effective manner always exists. It is claimed that importing high technology defence equipment means the loss of a capability and recreating this technology for future weapons systems development could be both time consuming and costly. But this argument ignores uncertainty and the nature of technical progress. Revolutionary rather than evolutionary technical progress could render obsolescent today's successful design teams. Moreover, in the future, it might still be cheaper to buy from abroad if we do not succeed in creating a market for our defence products. The challenge for the future is for the defence industry to develop and maintain its innovativeness and technological superiority while shifting towards a greater emphasis on higher quality and lower costs.

(c) **Can not be held to ransom**. Foreign supply leaves the buyer vulnerable to monopoly price increases. It is argued that once a nation is locked into a foreign supplier, perhaps as a result of a low price for the initial equipment the supplier is then able to charge monopoly prices for spares and support so that life-cycle costs are higher than the domestic alternative. Lifecycle costing should be undertaken for both home and overseas purchases. Otherwise a nation could equally well become dependent on a small-scale inefficient domestic monopoly providing unsuitable equipment.

(d) **Exact need may not be met.** Foreign supply often provides unsuitable equipment not tailored to a nation's requirements. There are enough instances where we had to set up modification facilities in respect of equipment procured from "much colder" East European countries. We even get saddled with items like "Scania Vehicles" which have limited practical utility in so far as we are concerned because these have been designed for colder climates and much heavier artillery deployment where NATO forces expected to fight. It does not follow that a national DIB will always provide suitable equipment at an affordable price within the required time-frame: but, at the very least, we will get a product which suits our conditions.

(e) **Leverage**. A DIB may enable a country to be a more informed buyer and improve its bargaining power when considering buying from abroad. The strength of the leverage will depend on the potential cost of domestic development, and the level of technological capabilities. Potential suppliers knowing that with some effort we could have an indigenous product in all probability quote a price which is affordable.

(f) **Economic Benefits**. DIB provides national economic benefits. These take the form of jobs, technology, and support for the balance of payments and a contribution to the national exchequer. (It is understood that in UK when comparing British and foreign equipment, a premium of 25 percent to 40 percent is applied to the overseas purchase price to allow for some of these wider national economic benefits).

The DIB in developing countries

Some of the benefits outlined above apply to the case for creating a DIB in developing countries, although they will differ in importance and relevance. A developing country might view a DIB as an important contributor to saving and/or earning valuable foreign exchange. Also, a number of developing countries regard the defence industry as a leading sector providing spin-off benefits to the rest of the economy. On this view, military R&D is seen as beneficial for the development of key high-technology sectors which, it is believed, through various spin-offs will eventually improve the international competitiveness of the economy. An important question however will be- whether the defence industry is more or less useful than other industries as a driving force for technological modernisation and a source of economic and infrastructural spin-offs in a developing country.[5] In this, a country like India will have to make deliberate choice.

An effort has been made to put across that contrary to popular belief 'Defence does contribute materially towards Development'. Obviously, this also implies that we manage the defence expenditure in a manner that ensures both; i.e., national security and growth of the national resources in the form of better life for the common man, concurrently. That is to say, we need a total integration of defence and developmental planning that, in turn, can lead to economics of scale in regard to production of high-tech materials and equipment thereby obtaining handsome returns on the defence outlay. The list of conceivable benefits can differ between developed and less developed countries. First, both types of countries can experience a stimulative effect from defence expenditures during periods of unemployment, caused by under consumption or under investment. This stimulative benefit effect, of course, ends once the economy is at or near full employment. In our case we will need too long a time to arrive at this situation. We are at that stage of development where we can experience direct technological effects and spin-offs from the defence sector, specifically if we adopt the philosophy that when the spin-offs from the defence industry are applied to the civilian sector, growth is promoted. Defence expenditure can effectively enhance growth if some of the expenditure is used to provide social infrastructure(e.g., dams, roads, airports, communication networks) and other forms of public goods.

A national, modern, and broad based re-industrialisation has to be undertaken so as to create a solid foundation for defence and defence related production. This consideration along with the need for self-reliance necessitates an urgent study of our industrial capabilities both in the public and private sectors. This also means the application of rigorous economic analysis so as to ensure that the resources are allocated efficiently among competing goals in both the defence and civilian sectors in order to achieve the dual aim of an efficient defence and a happy population. We in India have not so far made any effort to harmonise our defence needs with the civil sector. Advancing technologies with both civil and military usage can only be inducted with speed if we allow our industry to participate with vigour. Even when technological spin-offs can come from either the civilian or defence sector, the application of technological breakthroughs to the private sector is often faster and more direct if they originate there and are brought into the defence arena on purely 'need based' considerations. This can only be done if there is total integration of defence and civilian usage in as wide a field as is conceivably feasible. What it implies is that the capability of the nation to deter all kinds of aggression is to a very large extent measured by its ability to mobilise and to employ its economic and industrial resources in a manner which ensures sustained technological and material growth. Our policy makers need to examine the whole philosophy of "economic interdependence". A critical analysis of the population's prosperity and defence preparedness of those countries who have done so, indicates that considerable all round economy can be achieved if we identify those of our military needs which can be met by the trade. Thereafter, under the charge of 'Planning Commission' or equally powerful governing body we should identify those sectors of the industry that can meet our needs. A broad mutually acceptable QR can be prescribed for both the defence and civil industrial sectors.

To achieve its policy of upliftment of the poor and strategic objectives, the government must promote a well-defined prosperous and competitive economy as well as ensure the defence industry's continued access to foreign markets. We only have to harmonise the situation in a way that it ensures both, i.e. development and defence. And, if we agree with Pandit Nehru and those economists and defence experts whose opinions have been recorded above then whatever a nation spends on defence should logically and largely be considered as an investment. To put the whole thing in a proper perspective there is a need to analyse and to study the subject in totality, so that we identify the areas of mismatch in order to ensure that the equation so thoughtfully defined by our first prime minister is put into its appropriate developmental and defence-related environment. This means that today we need to make long-range technological assessments so that national allocation decisions for development and defence can be taken in a manner that is capable of actualising the common man's dream as well as provide the country with a viable defence. But in the kind of environment that we live, this will invariably lead to a situation where a large number of interests will clash. But that difference of opinion is not enough reason for us not to draw the above listed benefits by amalgamating our defence needs with civil needs. The expenditure on research, on maintaining large vehicle fleet by services and inventory carrying costs could be considerably reduced by integrating the two in a manner which guarantees a higher level of assurance for the military and improves the quality of life of the common man. We must realise that a very wide range of goods and services are either under utilised or are not taken into use at all in the civil sector. By making them available for defence and investing in defence oriented infrastructure we can actually ensure greater speed and efficiency in utilising these national resources.

Notes and References

1. India's Ad hoc Arsenal Direction or Drift in Defence Policy? by Charles Smith, *Sipri*.
2. *Affording Defence*, by Jacques S. Gansler.
3. *The Defence of the Realm in the 80s*, by Dan Smith.
4. The Mobilisation base is that capacity available to permit rapid expansion of production, sufficient to meet military, war-supporting, essential civilian, and export requirements in the event of a full scale war. It includes such elements as essential services, food, raw materials, facilities, production equipment, organisation and manpower. Implicit in the mobilisation base concept is the idea of conversion of industrial capacity from commercial production to military production.
5. *The Economics of Defence*, by Todd Sandler.

6

Economics of Logistics

Economics of defence, and therefore military logistics was born when man first of all armed himself with specifically selected sticks and stones. Earlier on the defence equipment was rather simple to make and maintain. Once produced, the sword, the spear or bow etc lasted for a long time, required very little maintenance and, if desired, could be replaced at any time with comparative ease. Accordingly, the impact on the economy was also somewhat limited. In any case, the victor after a war had total control on the assets of the defeated and the booty more than compensated for the investments made in acquiring the implements of war. Thus, there was no reason to actually analyse or restrict the expenditure on defensive or offensive capabilities. Defence effectively meant hardly any burden on the common man. It is only in later years, with the industrial revolution and with the invention of gunpowder the implements of war started getting expensive, and more and more complicated. The rulers became ambitious and keen on grabbing more territory. In order to protect 'trade routes' and territorial gains and integrity of the country a need for a particularly sized force that could be maintained appropriately was consequently felt. This meant an assessment of "defence expenditure" so as to gauge pros and cons of specific military adventures. This in turn meant that a deliberate economic and industrial plan was now required to be drawn for equipping the military within the limits of economic viability if prosperity was to be ensured. Prosperity needs peace, and peace calls for deterrence, deterrence needs an effective defence capability. The level of effectiveness is related to the amount of resources that we are willing to spend in equipping our military.

Of all the subjects to which economics would have a natural application, it is remarkable that "defence" has been neglected almost entirely, especially in India. The ignored fact is that there is an important relationship between economic strength and logistics capacity and capability. A nation's economic base generates the wealth that can be transformed into instruments for the execution of its strategic initiatives. It is our economic position both in peace and in war "that" limits the size of the forces we can create. Economics over the years has played the role of "revolving door" in both the military and civil sectors because we have been shy of treating defence as a tool of national development. For a developing country like India no aspect of military logistics- be it industry, economics, development, procedures & policies- can be discussed in isolation. Total defence calls for the coordination of military defence with the civil, economic, technology, industry and psychological components of a nation. There is no reason for a country like India, with its fairly viable, stable and advanced industrial base, its defence and civil production, research facilities, its experience and its considerable domestic demand, not to have a common forum dealing with the national needs in totality. This may even help in speeding up the indigenisation process and reducing defence reliance on imports where spares etc is concerned. There is a definite need for optimal utilisation of resources. For this to happen we need to appreciate that industrial power, economic strength and national security are inseparable. Logically, a logistics system should entail a national level perspective and direction. Then only we can provide a cost effective, efficient, reliable and dependable material support to our military. There, however, has to be a fine balance between the two requirements. Too much preponderance towards defence oriented technology with limited civil end usage, and planning production with only the military need in mind can be counter productive. (In certain quarters it is being said that one of the reasons for the collapse of the "Soviet Union" was it's over-indulgence in producing military hardware). Commercial viability of a product or technology will be an important factor in ensuring a successful policy on the subject.

Logistics from the point of view of an economist, may be said to depend on three things: (1) the quantity of national resources available, now and in the future; (2) the proportion of these resources allocated to national security purposes; and (3) the efficiency with which the resources so allocated are used. Economic strength of a country serves as the basis from which all logistics activity of the military proceeds. This is the foundation on which government and military leaders determine the ability of existing logistic assets to support a military strategy. In case they don't, then resources have to be found to match that strategy or the strategy itself is required to be modified. This means that the economic capabilities of a country limit the combat forces that can be created. At the same time, logistics capabilities limit the forces which can be employed in combat operations, for the forces to be ready for combat there is a need to create and deploy resources both in terms of manpower and material much before a combat like situation arises. A country's capability to divert these resources from traditional developmental activities and sustenance of the population determines the level of defence preparedness it wishes to have. If the national economy is unwilling or incapable of sustaining the defence forces created to substantiate a strategy, then the economic infrastructure must either measure up or that strategy be modified so as to fall within the capability of the country's economy. It is obvious that economic-logistic factors determine the limits of strategy, the size of military and long term goals of a nation in so far as its will to be given due respect by immediate neighbours in particular, and by others in general, is concerned. This also means that a country needs to have a strong industrial base, a higher level of prosperity and a capability to mobilise resources for measuring up to the national strategy. The economic act of industrial mobilisation is related to the grand strategy. Strategic posture of a nation is influenced by its economic capability to maintain and sustain such a posture. The operational logistic action is related to specific strategic plans.

All the classical economists have always held a view that the economic systems need little tinkering by the government and that too only in the fields of provision of education, defence, a judicial system and infrastructure. Certain well known economists have also stipulated that logistics is a synthesis of several areas of economics and administration. No single branch of government, in itself, can effectively create an environment where the strategic goals and material compulsions of the forces can be matched and met, unless we set national goals in that direction. In our country there has always been an opinion that the resources allocated to the military could have had a better utilisation. While lambasting the defence expenditure we tend to overlook the gains to the nation, in the form of trained and disciplined manpower coming to the rescue of the populace. Be it floods, earthquakes, and various calamities be they natural or manmade. The 'self styled' do gooders either ignore or pretend to miss the advantages a country has when its military is well equipped and well trained. They seemingly, overlook the fact that a strong and vibrant economy needs adequate and credible defence capability. This capability is the base for wide ranging security issues including internal security and social stability. Such security, in turn, is the base for savings, investment and growth.

Defence equipment, especially high technology equipment, is costly and the trends are towards rising costs in real terms. This means a greater strain on the country's economy. This also means that the demand for military expenditures is an important consideration in defence economics, because this demand relationship indicates the manner in which the country allocates its resources between defence and non defence goods (i.e., the classic guns verses butter trade-off). It is appropriate to examine the military logistics infrastructure and how to dovetail the same into the ambitions of a growing economy so that we create resources for the military needs with minimum and acceptable strain on the civilian economy, as such. No treatise on the subject of logistics can, in any case, be undertaken unless economic viability of what the military wants- in order to be an effective deterrent- and what the country can provide is also studied. There is no simple or generally agreed-on method to measure the "burden" of defence on the national economy. Much less, any economic ceiling that can really be imposed on numerous on-going R &D projects targeted towards self-reliance and building the national strength, knowing the kind of environment that we live in.

Adam Smith, the father of modern economics, justified expense on defence as a first duty of the government. He wrote in 1776, " the first duty of the sovereign, that of protecting the society from the violence and invasion of other independent societies, can be performed only by means of a military force. But the expense both of preparing this military force in time of peace, and employing them in times of war, is very different in the different states of society, in the different periods of improvement." Smith concluded that civilised nations would rely on a militia for defence at their peril. They must instead establish, and pay for, a standing army. He drew attention to the problem of the increasing cost of defence. In his own words," the great change introduced in the art of war by the invention of fire arms, has enhanced still further both the expense of exercising and disciplining any particular number of soldiers in time of peace, and that of employing them in time of war." In his view, it was evident that defence was for the common good and on this basis it was reasonable that the expenses of that defence should be met by the whole society with all the different members contributing as nearly as possible in proportion to their respective abilities. This means that the government is obliged to increase its activities in the field of policing and defence. Second, social welfare spending will rise with increased development since aspirations of the community will grow with the rise in national income. Third, the government may need to participate more fully in the production process owing to technological reasons. This will enable the industry to produce superior goods whose demand rises with income. However, if the military aspects of industry and developmental expenditures are given greater priority, and the economic considerations are only marginally taken into account, they may involve a miscalculation of resources leading to gross inefficiency both in military and civil sectors thereby making the whole exercise a 'loss making' proposal. One needs to put the whole subject in a meaningful perspective. Total involvement of the planners, industry and military for a combined and collective look towards the defence needs vis-à-vis availability and capability in so far as national resources are concerned can pay rich dividends in the form of economy and higher level of assurance.

Since every country likes to have a competent defence it has to find means and resources for acquiring that capability. There are the desirable things that we should be doing from the point of national security, but they may far exceed our economic capabilities. Certain hard choices will have to be made. But the greater our economic strength, the more desirable things we can do, and we can do them better. All it needs is a balanced view of the social, political, environmental and cultural scenario. Then only the resources in hand can be pragmatically allocated to various important commitments. The nation has certain resources that are conventionally classified by economists as various sorts of land, labour both skilled and unskilled, industry and capital. These resources are to be used to satisfy many objectives of the nation and the aspirations of its individual citizens- national security, a high standard of living, social security, a rapid rate of economic growth, and so on. In general, the more a nation spends on security the less it will have for upliftment of its citizens. Economics is concerned with allocating resources-choosing doctrines and techniques - so as to get the most out of available resources. This means study of various trade offs between defence expenditure and other forms of government expenditure. The trade-off between defence expenditure and other components of GNP[1] affecting the growth rate of the economy, especially savings- investment ratios. The impact of resource constraints like foreign assistance, fiscal deficit, adverse export performance and their adverse effect on the economy due to increased defence burden will be another arena for this purpose.

A large number of economists the world over, before and after Keynes till Benoit have held the view that defence expenditure is non-developmental in nature. Certain economists have said that defence expenditure may hurt or stimulate economic growth directly or indirectly because defence is a 'macro-economic' State activity. Emile Benoit uncovered a net positive association between defence spending and economic growth for forty-four less developed countries (LDCs) during the 1950-65 period. He identified a considerable number of things that have favourable effects of defence spending. He, further said that the defence sector contributes towards manpower training, supplies basic infrastructure, undertakes socially useful research and development, provides a source of employment and may undertake 'civic action'[2] programmes, all in addition to maintaining public order and security. He argued that if defence is growth promoting, then the share of defence devoted to GDP could be used as a policy instrument for inducing development. If, on the other hand, defence is growth inhibiting, then a reallocation of resources from defence to civilian uses may be conducive to growth and development. Conventional wisdom, (he continues) however, argues that defence commitment on the scale of superpowers during the cold war hampers growth owing to defence expenditures crowding out private and public investment. Today, that kind of defence spending could be taken as an aberration rather than a realistic analysis because it was more of an exercise in one up man-ship rather than a practical approach. Naturally both sides sunk in a lot of money and other resources in grandiose plans and an 'arms race' which no body could really win. We in India have had no reason to either think of or indulge in that kind of defence spending. We couldn't afford it either. An endeavour to have more armament than every body else in the region of influence is an expensive thought because it would mean more money spent in an exercise which may or may not give appropriate return. It will be improper to say that a total focus on defence will not be damaging to the national growth.

Expenditures on military research and development, it is said, might produce new knowledge, techniques and materials which may have direct civilian applications, and the use of this new technology would clearly increase the economy's productive potential. There is another point of view that the provision of defence by any country represents an economic (or 'opportunity') cost to that country. In that, resources so expended cannot then be made available for the production of other commodities. The subject of how much defence spending is enough is too wide and debatable depending on one's perception of "national strength" and the military's contribution towards that strength. There are, in general, two ways in which the problem of balancing defence needs against fiscal constraints can be approached. One way is to ascertain essential defence needs and then see if funds can be made available to meet them. The other is to predetermine, as a matter of fiscal policy, a rupee limit for defence expenditure; and thereafter refuse to satisfy any defence need that cannot be compressed within that limit. However, one cannot properly draw up defence plans on the basis of either cost alone or needs alone. The question should be how much is needed for defence more than it is needed for other purposes.

Impact on Development

The premise that expenditure on defence build-up does not contribute to economic development has dominated the thinking of development planners and agencies including the World Bank, IMF as also the intelligentsia, including in India. Yet such resources may not be entirely diverted from other uses in practice. For all one knows, some or all of the resources so used might have remained unemployed but for defence needs. The fact of the matter is that considerable un-utilised or under utilised capacity has always existed in the economy and put to constructive use vis-à-vis defence, handsome gains can be made in the form of overall development. Even where resources utilised by defence have to be diverted from other sectors, the value of the resultant output does not necessarily measure the value of the output diverted from the civilian sectors. Considerable development, both in the infrastructure and services can be made when a defence requirement is enmeshed with the development needs of the country. Defence expenditure can promote growth by providing nutrition, training, and education to a segment of the population. These human capital-enhancing activities will later on impact the civil sector in a positive manner because a trained pool of manpower will be available to other sectors of the national economy. The defence industry provides employment and training opportunities for local labour, which will naturally produce a beneficial economic effect. Defence expenditures do indirectly support growth by maintaining internal and external security. Since a secure environment is conducive to promoting market exchange opportunities and attracts foreign investment, the money spent on defence tends to come back into the nation's economy in some form or the other. Often the inflow of foreign capital can provide the transfer of advanced technology. It can happen that defence activities may have a growth inhibiting influence, if the activity and technology is confined to creation of armaments only. In such a situation defence may divert resources away from public and private investment that are more growth promoting. A total view of the military and community needs will have to be taken to avoid such a pitfall.

In a country like India the environment is not yet conducive to total attention towards either defence or development (inspite of what our leaders have been telling us for years) because we haven't accepted the fact that defence and development are intimately related. The problem becomes even more acute when certain "well informed" experts opine that our R&D and other resources which could be alleviating poverty and thus could be more suitably and fruitfully applied to civilian applications, at times, tend to get diverted to defence. But this can happen only if the R&D are confined to the military sphere. In the case of R&D related to medicine and consumer related technology defence needs definitely give impetus for speedy induction of technology in the civilian market. If we continue to import much of our arms, then defence activities can lead to an adverse balance of payments. That, in turn, will have a negative impact on potentially growth promoting capital inflows, thus making the exercise counter-productive. To guard against such a situation we should take policy decision at the highest level to quickly identify our capabilities both in the civil and defence sectors and go in for indigenisation thereby reducing our dependency on outside sources. Finally, the defence sector may limit growth through inefficient bureaucracies, delayed decision making, and excess burden created by taxes used to finance the military expenditures which do not take a total view of our own capabilities that exist or can be created either in the public or private sector. To safeguard against such a situation, an organisation like "National Security Council" can be set up. This council will be in a position to take the total view of the situation and make allocations.

Economics of the Defence Burden

Defence expenditure primarily includes manpower, weapons, equipment, munitions, and spare parts for upkeep and maintenance. The manpower complement, its salaries, clothing etc. consume a lot of money and one cannot visualise any direct reduction unless the manpower in itself is reduced. Keeping in perspective the internal and external compulsions of today that does not seem to be a possibility in the near future. The environment that we live in indicates that there is no choice but to bear all this burden of the 'defence expenditure'. For, experience both in our country and elsewhere in the world has shown that (adhoc) cuts in defence expenditure affect preparedness and cause disruption in much higher proportion than the cut itself. A major factor that takes away a huge chunk of the defence budget i.e. stores, maintenance complements like spares, medical and health services and facilities like transportation etc., are some of the facets that, if brought into a common ambit at national level could help in identifying redundancies and, thereby bring in economy of effort and resources without jeopardising security. This would need examination at the highest level of "policy making". (The concept of a ' national logistics grid' is being discussed separately in this treatise). But, this will need a lot of national spirit and a will to make adjustments and sacrifices. Here again we need to formulate a national policy wherein the responsibility of various groups and organisations should be specifically defined with regard to what they have to do in case of a national emergency. This will help in ensuring that unnecessary controls and overreaction of certain individuals does not create an environment of panic leading to a 'ripple effect' where common man is concerned.

In our country (in all countries for that matter) the defence budget consumes a considerable portion of national resources. Not unreasonably, a question has often been raised, as to, how much 'national-security spending' the economy can afford. The allied concern is that short-run considerations may impair the long-term capability of the economy to support a large and sometimes expanding array of defence-oriented plans, needs and programmes. Defence expenditure has grown from Rs 10,477.45 crores in 1987-87 to Rs 55,500 crores in 1995-96 budget, that is, an almost 2.5 fold increase. However, if we take the inflation rate and the falling value of the rupee into account, the actual expenditure (in terms of Rupee value of 1980-81) has been more or less constant around Rs 6500 crores over the years. The real purchasing power is even less if we consider the higher inflation rate in defence equipment and devaluation of the rupee. The "Economy Advisory Council" has repeatedly drawn the government's attention to the rise in defence budget in the recent years. (That the rise is hardly enough to even cover the inflation and that the defence budget in real terms has actually being reduced has always been discussed in various forums with a sense of dismay). In order to make the observations more effective, the advisory council has also highlighted that the ever-increasing foreign exchange contents of the defence budget is a 'scarce resource'. The problem is that money in terms of Dollars for capital expenditure has dropped to 1/3 its value over the past few years. Our need for technological upgradation induction of better and more sophisticated equipment in all the three services has created a tremendous strain on our foreign exchange resources. Some leading economists in India have even opined that the foreign exchange crisis of 1991-1992 was precipitated due to earlier heavy defence imports that had to be paid for in hard currency.

Flexibility is one of the greatest assets of a free economy. We must be wary lest in our enthusiasm for industrial economic controls which we may need in time of war, we lose the flexibility produced by individual initiative. We also have to remember that whereas economy is essential to the attainment of the military objective, economy in itself, is not the objective. Therefore, all measures affecting the control, development and/ or induction of new equipment, and the coordination of the defence logistics must be judged by their impact on sustained combat effectiveness under war conditions rather than by the sole criterion of peacetime economy. By wanting to economise on the military equipment which is considered to be essential we may create a situation that is not likely to be acceptable to our population. I think if the Services could put across the valid theory that a repetition of 1962 could cause a permanent dent to the national psyche; India's interests will be well served. For a country like India with so many threats to the country's integrity economising only involves deciding how much of other things to sacrifice in the interests of enhancing our military strength. Secondly, being truly economical does not mean scrimping- reducing expenditures no matter how important are the things to be bought. Nor does it mean implementing some stated doctrine regardless of the cost. We should, rather be concerned with allocating resources choosing doctrines and techniques-so as to get the most out of our available potential and the capabilities that we can create in a planned and time bound manner. Where the Military is concerned it should be possible to bring home the point that an economically efficient solution to military problems does not imply a "cheap force" or a small military budget. It simply implies that whatever the military budget the country can afford the greatest military capabilities are developed. In the past, when battles were fought with less sophisticated weapons and distinct lines separated the opposing forces, it was easier to have a well-defined rear where combat service support could be performed with relative ease. Today, highly mobile forces will operate in an extremely hostile environment. These forces will use, and the logisticians will be called upon to support, weapons of unprecedented complexity and lethality. It will require anticipation, a different kind of training, integration, continuity, responsiveness and improvisation in a big way. All this also means that a huge amount of money will be required firstly to equip the forces and then to maintain and sustain the costly equipment in the hands of the troops. Thus, modern technology presents Army Logisticians with both opportunities and challenges. The main problem, however, will always be to find enough money to create a strong defence force.

Our economic position both in peace and in war limits the size of the forces we can create. Further, logistic considerations inevitably will limit the size of the forces that we can sustain and maintain. More so, when we have to maintain troops in all kinds of environments and terrain throughout the length and breadth of our country. Therefore, the economy achieved in logistics, the greater will be the effectiveness that the combat forces can develop within these basic limitations. There will, however, always be a need to evaluate the influence of any proposed step whether economical or political on combat effectiveness once the national defence gets associated with a particular proposal. We need to understand that the defence logistics is directly related to the country's technological base, economy and its industrial clime. We have to meet our defence requirements in a manner that allows growth of both, the military and the civil. This means some kind of sacrifice will have to be made till such time our industry and economy catches up with the advanced countries. Every country has gone through this phase. The operative clause is that a country has to be rich in its heritage, its tradition of self-respect and esteem to be able to stand up to those problems that are manifesting themselves in the form of numerous jabs at the country's security. We must have a will to make sacrifices for the country's well being. A population's commitment to a strong defence goes a long way in providing an environment wherein the government of the day could expect its populace to willingly and cheerfully accept its burden for an important cause like the national defence. It will be important to accept the fact that economics of a viable defence is an essential input in order to study the logistic support required by the military in today's environment. Economic constraints should, however, not be the most important limitation on the level of material support.

One factor which always has to be kept in perspective is that the division of resources between the military and the civilian economy is a question so vital to the welfare of the entire nation that it must always be answered at the highest policy making level. Optimisation of competing demands for resources can only be done at highest level because it is only at that level that a total picture of the national resources, the vision for national growth and a total focus of the world situation and what role we assign to ourselves, is available. More importantly, the strategic vision will help in assigning priorities. Areas of commonality between development needs and defence can also be easily identified. This will help in selecting those areas where development- be it industrial, technological, material or improvement of human resource- is required to be undertaken. Depending on factors like urgency, political and social considerations the priority can be assigned. There is an urgent need for effective coordination in evolving production policies and their implementation in the administrative decision-making echelons of the defence ministry, other important ministries, Ordnance and other production units. Proper production planning and efficient operations of these units cannot but result in worthwhile economics in production, as well as, saving in materials and energy. As our economy becomes more complex and as our military requirements include more and more of the products of industry, the requirement of "management information" and management of that information becomes more and more problematic and intricate. This calls for mechanised and electronic reporting and processing systems at a structured level of decision making. This aspect will be discussed and at some length a little later in this narrative. In the ultimate analysis the logistic system has to be in harmony with its supporting economic system and the country's industrial base.

Notes and References

1. Gross national product, or GNP, is simply the rupee value of a nation's final output over a period of one year. It is the total volume of goods and services, valued at market price, that is at the nation's disposal over the period. Unlike net national income, GNP represents the whole of the nation's output, no portion being already set aside to allow for depreciation, therefore, any consumption, replacement of wornout or obsolete equipment, additions to the stock of capital military outlays, or other government expenditures must come from GNP.
2. Here, the term 'civic action' describes cases where the military provides deliberate benefits to the civil economy by acting in a 'non-military manner'. Our troops have been doing so by providing "medical and health" facilities in remote areas, besides any number of other socially useful projects. Besides, troops and families spend their money at the place of their posting, thus adding to local economy.

7

Bureaucracy and Logistics

Show me a case and I will show you the rule.

Way back in late 19th century one Captain Kennedy while vacationing in the hills found a place that he felt was just right for the 'summer capital' of the Raj. That is how the hill station of Shimla came about. Today, even for siting a small logistical or any other unit (like an ammunition sub depot or an engineer's platoon) a board of officers will be constituted. Headed probably by a brigadier or a major general. He will have members from the directorate of estates, some special adviser from the Ministry of Defence, a member from the environment directorate, representatives of various wings of the state government etc. These advisers will be giving their opinions about the lay of the land, climate, weather, suitability of the location and so forth. The case will move, if at all, through the 'staff channels' of the Military for years (original financial estimate would have become redundant in the meanwhile). After one has overcome all the objections and uncertainties of the Military hierarchy the file will be moved for, what is termed as, "the Government sanction". It will normally be shot down by the bureaucrats for one reason or the other till either some desk officer or any body at equally lower rung takes mercy and moves the file for somebody's approval as what is termed "as a special case". In the mean while the state government (in whose area of administration the unit is proposed to be located) might either have had a change of heart. Or there could be a different government in place who would disagree (on the principle that the rival party's decision need not be upheld) with its predecessor. Assuming one has been able to overcome all these hurdles in about five to ten years time there will be no funds available for the project, which in any case had by now become costlier because of "escalation factor". Again one goes to the Ministry. This time to the Financial Advisors, who being busy with overseeing defence purchases of all kinds, (right from underwear for the soldier to the purchase of aircrafts and tanks) will normally have no time for such a complicated subject like acquisition of land. (Our 'Land' laws being what they are, this is not really surprising.) Building of sheds for storage of ammunition, or accommodation for the troops will continue to suffer in the meanwhile. All this time the concerned 'Service Head' will be giving long winding explanations to the General Staff vis-a-vis his inability to meet an operational need in a particular theater. Eventually the project may or may not see the light of the day. Based on lifelong experience the bet is, it won't- unless some bureaucrat or a politician is personally interested in seeing the thing through. When a bureaucratic system works well, it encourages people to perform the best they can, and to contribute to the institution in ways that can be assimilated. Napoleon obviously knew this, and that's why he invented the bureaucracy. Despite the evidence of the existence of sophisticated machinery designed to facilitate a constitutional and rational decision-making process in India, the system does not work well in the realm of defence. May be it is so because democracy at work is always a complex and dynamic mix of power, opportunism, bureaucratic politics and political pressures

This is what the first Chief of the Naval Staff had to say about the bureaucracy when the 'Services' had not become as weak as they are today- "wise counsel ... helped me to exercise restraint in periods of frustration(unfortunately the dismissed Naval Chief did not possess this trait). These frustrations arose chiefly from the bureaucratic machinery.

Bureaucrats fall into two categories. There were those who knew all about every thing, including operational and technical matters, and those particularly of the Finance Ministry, who did not seem to care what harm they did to the Service so long as they saved money for the exchequer. There was a third neuter group whose effective contribution was minimal. The basic fault lay in the system of functioning of the ministry whose officials played no part in the initial formulation of plans, thus depriving themselves of the opportunity to appreciate both the professional considerations and requirements as well as financial and practical limitations that are involved in any proposal. They preferred to remain the ultimate arbiter."[1] (The recently dismissed Naval Chief did no good to his cause when he said," It suits the civil services to have the public at large, and politicians in particular to believe that the military mind is somewhat dull, hidebound, hawkish, inflexible and , last but not the least, ignorant of subtleties of issues other than the military. They shy away from the suggestion that it is the generalist administration which ceases to grow intellectually since they undergo very little formalised training, and the major portion of their time is devoted to the technicalities of rules and the dictates of politicians.") Fact of the matter is that the bureaucrats take great pleasure and draw considerable satisfaction from diluting, denying, delaying all requirements which the military considers to be essential. Military is a complex organisation. And, complex organisations need structure, which means bureaucracy.

Some where it has been mentioned that, "Bureaucracies the world over display a major commitment to self-maintenance or expansion. This commitment works both at an individual level, due to preoccupation with job interest or security or promotion prospects, and at an organisational level, due to more collective concerns with the survival of the organisation as a whole. The content influence manifests itself in a host of ways. For example, it has been argued that bureaucracy has a propensity for low-risk policies in order to decrease the probability of failure, or for policies that entail rapid expenditure, in order that future budgets can be justified. Bureaucrats are more occupied with the initiation and expenditure on projects rather than the impact of the projects."

A noted bureaucrat Mr S.K Das in his book "Civil Service Reform and Structural Adjustment has observed, "Quote"... " Civil servants, as maximisers of their self-interest, are also interested in economic gains and in pursuit of their careers. While some civil servants may selflessly pursue public good, most officers do so while simultaneously pursuing their careers and economic gains. The bureaucracy thus, has an entrenched interest in continuing with inefficient policies."[2] Unquote, over the years in our country for whatever reason the role of the bureaucracy has become that of a watchman over the military rather than being supportive. What is even more disturbing is the truly trivial, parochial, and intrusive activities that masquerade as legitimate "overseeing" functions. No one can credibly contend that all these actions will result in a better defence-preparedness. This continuous control by those who are not accountable to the nation (regarding security) or to the soldier for his inadequate equipment bring instability, uncertainty, inefficiency, and reduced effectiveness for the taxpayer's money. To compound the problems of the "logisticians" the world over there seems to be an inevitable tendency in bureaucracies for decisions to be made at higher and higher levels. There are some good reasons for making decisions at high levels. A broader and less parochial view can be taken. But on the other hand there are excellent reasons for making most decisions at lower levels. The man on the spot has intimate first hand knowledge of many factors relevant to his decisions. Most other nations of the world have had leaders who have been through the grind or who have seen active soldiering as a part of their 'growing up'. In their environment it is not difficult for the Armed Forces to push their genuine cases through. In more advanced countries bureaucracy usually has military experience. The equation is, therefore, of equals having adequate understanding of each other's problems and limitations. It is not as if the military has no problems with their bureaucrats or politicians. But the situation rarely goes out of hand. The problems exist for a country like India, where neither the bureaucracy nor the politicians have, generally, been anywhere near a scene of actual battle. (We rarely had a defence minister who knew his onions. Pandit Nehru made Mahavir Tyagi his defence minister because he had served as a "cook- non combatant un-enrolled" during the First World War and was therefore considered to be wise enough to deal with the country's security. There were notable exceptions like Mr. Arun Singh. Unfortunately, he did not last long enough to reign in the over exuberant babus who have been forever insulting the professional acumen of the men in uniform. It is understood that he often lamented the fact that the bureaucrats behaved more like "ministry against defence" rather than " the ministry of defence"). Consequently, the mandarins take most of the decisions in the cool comforts of their air-conditioned offices far removed from the ground reality. By default the Military suffers in the bargain.

It takes years and years for the files to move at various levels. In the meanwhile the prices keep escalating, the equipment in the hands of the troops keeps aging and getting out-dated leading to even greater problems of maintenance and up keep. But then the problem is to get, what is known as, the government approval even for small items that are essential for day to day operations and functioning. Sadly this government approval is also often given rather grudgingly and that too what the bureaucracy thinks should be adequate regardless of the impact such a decision can have on the troops safety, morale and well-being. The job of economising on the military budget has been given to budgeters and financial comptrollers who are not necessarily up to the task of making military decisions. In the opinion of Mr. H.M. Patel, a former Defence Secretary, the policy organisation of the Defence Ministry was 'sufficiently flexible to ensure that every relevant point of view has a chance of being presented at an appropriate level if necessary'. The theory is rarely if ever translated in practice, however... The ability of the average civilian to make such decisions... must be judged against Mr. Patel's own admission that the ignorance of civilian officials (to which may properly be added that of the politicians) is so complete as to be self-evident and incontrovertible fact. (Admiral Bhagwat was not very much off the mark when he made the observations listed above). Root cause of the whole problem seems to be the misplaced notion that the bureaucracy has of its intellectual superiority. The author hasn't felt that this 'superior' intellect exists exclusively in the bureaucrats during his numerous professional or personal dealings with them. Most of the military officers have as good or even better competence than the bureaucrats who take important decisions on what the military should buy and from whom it should buy.

Much of the present problems, logistical inadequacies, delays and inefficiency could be removed simply by making the 'Ministry of Defence' and 'Ministry of Finance' realise that they bear a part of the responsibility for seeing that the military does its job properly and in time. Where any thing is in doubt it should be settled by discussion among three equals and responsible bodies, rather than by the military being subjected to endless inquisitions to prove its case to the ministry. It is alarming that a case for "bullet proof jackets" has been lingering in the ministry for the last so many years (Times of India 9Sep, 1997). This, while the officers and men are dying at the hands of 'foreign sponsored' terrorists and insurgents who, more often than not, are much better equipped than our troops. Sepoy Bhup Singh's morale is usually low because he is not getting his uniform items in time. The bureaucratic control and its lack of sensitivity has been causing avoidable delays in purchase of even 'daily use' military items like blankets or even shoes. The troops are being denied their legitimate dues because concerned bureaucrat or any of his flunky may feel that granting a sanction without sufficient number of requests and pleas for action by the user the bureaucratic control over military is likely to be compromised. The adverse effect such delays have on the morale of the troops is difficult to define. The effect of the instant lack of governmental sensitivity is too far away for an average soldier to comprehend. For an average soldier it is not possible to understand, as to, why his superiors cannot get him the wherewithal for fighting, or even his "life cycled" clothing items that remain overdue for prolonged duration. All he knows is that his superiors have let him down. The Defence Minister sent some bureaucrats to the Siachen glacier to experience the things for themselves as they had been sitting on files pertaining to urgent requirements pertaining to that region for years. Hopefully, this becomes a regular feature for the bureaucrats to have first hand exposure and to experience the problems faced by the soldiers.

It is a common refrain in the ministry that the Military almost always overstates its requirements. Whether this is so or not is not the point at issue, but the fact that such mutual distrust is not doing any good to either the troops or the country. (One could never understand, as to, how and why only one fog-light per vehicle was sanctioned for the vehicles operating in areas like Tawang in the North-East. The fog that exists at places in that region even during a bright and sunny day and at eleven o'clock in the morning should really have not required any case to be made for the fog-lights. But it had to be done and the powers that be sanctioned the barest minimum that they felt would be enough for the safety and wellbeing of the jawans traveling on those winding and steep fog-engulfed roads. It was ostensibly done in the interest of economy.) If the bureaucrats or the politicians concerned had ever served in such areas and had traveled on those treacherous roads the troops would have had a more sensible deal. And the (not infrequent) loss of trained manpower and vehicles due to inadequate wherewithal could, perhaps, have been avoided. The only rationale behind such an irrational authorisation and no amends being made is the lack of military experience with those guardians of 'public money' who would have gone home happy that they had managed to save some pennies for the exchequer and have no ostensible concern with those troops who may be dying for want of proper facility. There is only one solution to save our troops from this kind of economy and judgmental bureaucratic control. Only those Civilian Officers be posted to the 'Ministry of Defence' who have had a tenure of, atleast, three years in the "valley" or in that part of North East where the anti-national elements are active or where the man is busy fighting the elements. May be an exposure to a real danger will bring in an element of urgency and realism when the Military moves its cases for specific equipment.

It needs to be accepted that in a democracy where basic control functions are performed by a bureaucracy the interplay of civilian and military in control of logistics will always involve overlapping areas of a variable nature in which men must meet, work together take decisions and share responsibilities in a spirit of mutual understanding and cooperation. These areas are not definable by law. If law attempts such a definition with a view to smoothen the present incompatibility between the military's and the bureaucratic thinking on the subject it will either be ignored, or be circumvented by the decision-makers informally even if any arbitrary and restricting legislative or executive order exists. The aim of the whole exercise should be to ensure that the troops are not denied their essentials because of a narrow or antagonistic approach of those concerned with according sanctions. In order to safeguard the interests of the troops accountability and responsibility must be well-defined for a system to work smoothly and constructively for it to provide meaningful controls and checks, so essential in a democratic set up. Unfortunately, over the years these checks and controls have been reduced to meaningless exercises where bureaucracy has abrogated to itself the role of government.

Through out service and whenever a case got stuck due to bureaucratic inertia one could never understand as to why is the bureaucrat considered to be the government and why a service officer who probably has more at stake where the country's security and safety are concerned is not taken as part of the government. A large number of bureaucrats themselves (often after retirement) feel that their service is not measuring up to the country's needs and its defence requirements. To get over this logjam we could 'laterally' induct retired senior service officers in the Ministry of Defence, in Pakistan the Defence Secretary is a retired Lieut. General. Having such an expertise in the ministry can help speed up the decision making and save those precious lives now being lost and speed up the procurements now being delayed because of the apathy of bureaucracy. Recently, the defence minister has undertaken an exercise in the direction where officers at the "joint secretary" level have been inducted in the ministry. Services will do well to sponsor those officers for these assignments who have experience of procuring equipment for the military. Else, the whole exercise could become meaningless and counter productive.

Another positive step is to allow the military to expend its revenue budget at its own discretion. Next logical step should be to let the military work out its own priorities and expenditure where overall allocations are concerned. Today, military has experts virtually in all walks of life, including finance and economics. Therefore it may not be an entirely a bad idea if the military draws its own budget, assigning priorities, monitoring all aspects of its allotments within the framework of the available resources. This kind of arrangement will also remove the wasteful duplications and separate procurements of the same items and from the same vendors- often at different rates- by the three Services. One eminent American economist (Professor Abba Lerner) has proposed that the whole military establishment should be organised in a network of markets and that it operates, by means of decentralised decision-making, much as the private economy does. The Theater (Army Commanders) is given their share of the budget to plan and equip forces in their areas of responsibility. According to his plan a theater commander would be permitted to make allocations of budget to his subordinate commanders "whenever he has sufficient confidence in the judgement of his subordinates and where their knowledge of the detailed needs of their particular requirement was greater than his own." We being a large hierarchical organisation tend to be sluggish and hide bound by rules and regulations. The proposal is far too unconventional and Service traditions far too strong to even merit a serious discussion in our environment, For that matter, even the Americans do not seem to have taken it seriously. The idea, however, has merits for limited applicability in certain spheres of our logistics needs. There could be a governing body consisting of representatives of the Ministries concerned so as to ensure that the things move smoothly. There may be the apprehension that, if military men draw up the military budgets, there will be a disregard of the over-all economic and social welfare of the nation as a whole. But that, to my mind, would be an entirely misplaced lack of faith in the commanders. All that is required to be done is for allocations to be made and adhered to so that plans are suitably spread out and funds appropriately utelised without delays and arbitrary reductions being made by the ministry as is the case today.

Uniformed Bureaucracy

To be fair to the Indian bureaucracy it will be appropriate to record some experiences with what is described as 'staff channel' and more appropriately the "proper channel". The officers posted to these channels (there are 'G', 'Q' and 'A' channels) generally are very vigorous in safeguarding the sanctity of their respective domains, particularly when they are posted to an Area or a Sub Area Headquarters. Usually, there are only logistics units located within their areas of influence and these units are too far wide spread to be under any kind of effective control as such.

There are thus very few chances for them to display their "obstructionist skills". Following case will illustrate the point. CAD at Pulgaon received a heavy consignment of ammunition by sea at Cochin from abroad. Communication from the unloading agency that they will unload the ships and demand railway wagons for transshipment only when an 'escort party' reported to them was received by the depot. It was also intimated that after a particular time span the shipping company would levy demurrage charges. Since Headquarters Command was to detail the escort party and the time was short the requirement was taken up with that headquarter directly by the depot keeping the intermediate sub area and area headquarters informed. Instead of detailing any escort party and saving on heavy demurrage charges that were likely to accrue all the three headquarters wanted an explanation from the depot commandant for bypassing appropriate authority. The explanations that due to urgency of the situation the depot had to take up with the ultimate agency that had the mandate to provide necessary resources for the purpose cut no ice. No body agreed with this contention and the case moved up and down. The depot eventually had to manage on its own because the second stage would have been to find some body to be blamed for the demurrage charges. There have been any number of such like cases where instead of smoothening the functioning of logistics installations the uniformed bureaucracy created situations where even the warranty periods elapsed because one or the other channel wanted to do thing in the classical bureaucratic fashion. A little bit of pragmatism and consideration for the limitations under which our service units functions can be a big help.

Notes and References

1. Kathari, Admiral R. D., *A Sailor Remembers.*
2. *Civil Services and Structural Adjustment*, by S. K. Das.
3. Abba P. Lerner, *Design for a Streamlined War Economy.*
4. *Financial power have since been revised. MGO has been made a "Budget Controller" in so far as the ordnance and EME budgets are concerned.*

8

Teeth to Tail Ratio

From Alexander to Giap in Vietnam there are a host of examples where logistics were the overriding factors in the success or failure of the operational plan. Those days are gone when Changez Khan and his tartars rode through a victorious ride almost all over the world. They went to places but never stayed, merely seizing what they immediately needed of food and clothes, and going on to the next objective carrying off gold and silver and whatever they could take along. In India, we had the 'Pindaries' who only had their horses, their weapons, some dried chappaties and a mission. Their need for logistics support was limited, as the weapon were their swords and horses the means of transport. Over time, soldier-centered resourcefulness shifted to a supply-oriented system in which the soldier became dependent upon a supply system instead of his own resourcefulness. Further, as armies shifted from extemporaneous to standing, preparation for their support became prudent. New considerations such as the locations of stocks started impinging upon strategy. Care and management of the materials started adding to the tail.

In the early history of man when the wars were fought each man had to find his own food, stones and knotted clubs. Each warrior was his own logistician. Not until later when fighting groups became larger was there any basis for designating certain men to specialise in providing food and weapons to the combatants. These men who provided support to the fighters constituted the first logistics organisations. With the passage of time warfare became more and more intricate. With science and advancing armament technology gaining control of the war making mediums, particularly after the invention of the 'gun-powder' and progress in means of transportation etc., over a long period of time, the need for expertise in various fields like medicine, material management et al kept growing. This naturally led to a situation where a much larger (and often cumbersome) complement of the support services got created, thus giving an impression of imbalance because these support services invariably seemed to be consuming too much of resources. They were seemingly denying commanders the desired level of mobility and flexibility, so essential to exploit their bold tactical concepts and plans. This often led to a situation where the commanders started feeling that the weight and size of the material support had, perhaps, become too unwieldy for short and swift operations that they envisaged. With that feeling of being cramped by the sundry service elements started the "Tale of the Vanishing Military Tail". Experts started analysing and theorising about methods of providing an efficient material support with as few 'service troops and technicians' as they felt were essential. In their hurry to find urgent solutions and means to make their combat forces lean and mean they forgot and continue to forget that ever evolving technology and arrival of more intricate systems meant greater involvement of a much larger complement of experts in an ever widening range. Increased technological developments have, thus, created further problems. Not the least of that is the need for rapid large scale industrial and training operations required to support the new inductions, and ordered up on short notice. There is no simple answer to this conflicting requirement of efficient support and a smaller tail at the same time. The fact of the matter is that the conduct and support of war now includes almost every activity and organisation in a nation. Most commanders assume that further logistical support from the "tail" is available. But the issue is not that simple. Warfare is no longer linear, meaning that there is no longer a front and rear to the battlefield. Communications allow control of fast-paced forces, consisting of army, navy and air force, over hundreds of miles. It can be said with a fair amount of certainty that in future wars, the forces, both the teeth and the tail, would operate in an arena saturated with second/third generation weapon systems and force multipliers with the electro-magnetic spectrum adding a fourth dimension to it. Service units in the combat areas and installations in the communication zones would be within the range of tactical missiles and attrition. A noted expert in the field of logistics has observed that through the changes marking the historical progress of logistics, one theme runs: the pressure of specialisation has increased the interdependence of elements of military forces. This trend is consistent with society's ever-increasing division of labour, moving farther and farther away from self-reliance. Today's fighting soldiers need all manner of specialised, high technology, support for their weapons and their persons. No one provides his own horse any more! With an ever-increasing demand for enhanced support to the front line soldier as far forward as possible and with an ever-decreasing manpower and financial resources, support services are finding it more and more difficult to measure up. In a world of high technology and limited resources, logistics need to be given due importance if we want to ensure operational success.

It has been universally experienced that more intricate and sophisticated our weapon and communication systems become, more is the number of technical personnel required for keeping the equipment 'battle worthy'. Thus, we are faced with a real dilemma of a need for creating a logistics organisation, which is both flexible and highly responsive to the operational requirements, both in peace and war, and is also compact enough to subscribe to the theory of a small tail. In addition to providing an efficient material support it should also ascribe to the current thinking where almost every commander worth his baton is busy finding ways to cut army's tail. They tend to overlook the fact that abstract speculations, theories, and principles where these concern 'Logistics' have never prepared a nation to fight and have never won a war. May be all that these have ever done is to enable man to understand his war problems, and, if he has been sensible to assist him in solving them. It is seldom appreciated that the intricacies of logistics requirements create a situation where a commander must understand the cause and effect relationships that exist in logistics. He must estimate how the gain or loss of efficiency in any particular technical function will influence the efficiency of the other functions that in combination determine his over-all combat effectiveness. It is only then an efficient and realistic 'tail to teeth' ratio can actually be prescribed.

Somehow, in our 'resource-crunch' environment a cut here or a trim there insofar as the Support Services are concerned has become some kind of an obsessive pass- time with our decision-makers. Since there is no hard and fast rule which describes an ideal ratio nor a mathematically defined formula to support such tail-cutting, this 'toning up' and trimming of tail by reducing odd tradesmen from various service installations could lead to serious problems in due course. It must be recognised that in the absence of a scientific approach such cuts tend to impose strains on the system of maintenance of the troops and equipment in their hands as the cuts are neither linked with improved support nor reduction of work-load of any kind. In earlier days when the material needs of the troops were limited and the equipment required hardly any technical expertise, an Alexander the Great could burn wagons carrying logistics elements because he was certain that the granaries of captured territories could support his forces. No commander can take that kind of chance today, nor can he dispense with his technical complement in the hope that he could capture enemy technicians and overcome his problems of equipment maintenance. A total and pragmatic view of the whole thing is called for because restoring a cut off portion of the military tale will be much more difficult.

Through out history one can find any number of instances where some commanders, in an obvious hurry for greater glory, took unimaginative measures to shed what they felt was an unnecessary encumbrance. That, they paid very heavy price for their foolhardiness has got recorded in some chronicles, but has hardly ever made any serious impact on the thinking of the combat commanders. Regrettably, an unanswered and rather intriguing facet is- that the field commanders, for whatever reason, generally did not (and often still do not) spend time and effort in understanding the complications and implications in providing logistics cover to the operational forces. They do not appreciate the fact that the logistic doctrines developed as a result of experiences gained in conventional warfare over the years are not necessarily applicable in today's environment when there are no well-defined rears and front. Weapons with better accuracy and higher rate of firing and much more lethal effect may call for a reduction in the strength of an infantry battalion or an armoured regiment. But these weapons with the technological proliferation need much more of technical and spare part support than the antiquated .303 rifle or the 25pdr gun, and therefore, much more support from the 'tail'.

According to Dr. Van Creveld "teeth to tail ratio" is frequently cited as a rough indicator of an army's efficiency. Low proportion indicates a high efficiency. However, many scholars have expressed doubts in expressing efficiency in this manner. The aim of a military organisation is not to make do with smaller number of supporting troops, but to produce an efficient and powerful fighting machine. History is replete with examples where this obsession with reducing support services and cutting of the tail led to creation of inefficient organisations that often proved disastrous. Whenever the discussion gets around to the tail and teeth syndrome I like to quote an old 'AESSOP fable' where a fox while trying to ape hunting skills of the tiger creates all sorts of contortions while jumping on a wild buffalo. Thoroughly mauled by the annoyed buffalo he sits down to analyse the reasons for this debacle. His eyes were as red as that of a tiger, his muscles were as tense as that of the tiger, and he had growled with equal ferocity, so what had gone wrong? "Well, the tiger before jumping on a prey, in addition to all that was copied by the fox, also tightens his tail, puts a lot of strength and spring in it. In short, he ensures that his tail is capable of supporting his spring, and that is something you did not have," observed an old fox who had closely watched the whole fiasco. We hear a lot of disparaging comments on the Army's tail so often because we tend to forget that a strong, healthy, vibrant, and appropriately sized tail is an important complement of any successful campaign.

It has become virtually fashionable to order studies and issue directives to the Services to dispense with this technical training or that functional operation, in order to provide officers and men for meeting the shortages in the combat Arms. Not for a moment is it realised that whereas an infantry company in a peace or non operational area can be handled by one or two officers, in an Ordnance depot or an ASC depot the stores are to be physically held, accounted for, preserved, maintained, and policed against pilferage and theft. These are also to be selected, inspected, packed, and accounted for when an 'issue' is to be made to the dependent units. All these activities may appear to be non-consequential to an operational commander. But these have to be performed and performed with efficiency. This means sufficient number of trained men and other resources. A fifteen years old ban on recruitment of the civilian staff, its aging and therefore declining efficiency of the available manpower has made officer level supervision more important than what it was a few years ago. The situation is further compounded by the fact that most of the storage sheds in our depots were either built in a hurry and with temporary structures during the 2nd World War, or again, as makeshift arrangements immediately after 1962. With their dilapidated storehouses having crumbling walls and unfit for deployment of modern "material handling equipment" the Service(s) installations today face tougher challenges than before. The installation commanders increasingly worry about facilities maintenance and serviceability of items on their charge; and quality of professional services they are required to provide, essential to retaining a quality force. Deterioration of facilities, decaying infrastructure and quality of life are critical concerns. With ever increasing "budgetary cuts, arbitrary ban on recruitment of the working level staff, deferral of facility repair and maintenance, reduction in station operations and engineering services and further, restrictions of utility usage are becoming common place. This unplanned and unrestricted reduction in the workforce is leading to a situation where the support structure may even give way some day. The situation is as grim where other Services are concerned. Be it a Supply depot or a Workshop, inadequacy of technical manpower, appropriate machinery, and modern work environment creates all kinds of problems.

The higher management does not seem to appreciate that creation of an environment of reduced material support and the so called, hard scale of officers in the 'Services', and reducing the establishment is counter productive. Both in the long and short run these are loss- making propositions. There already have been major fire-losses etc. attributed to causes like failure to 'look after' by the sub-unit commanders amongst other reasons like deficient "fire staff". To compound the problem of the Services the 'Defence Security Corps' (DSC) created exclusively of ex servicemen and to provide security to 'defence installations' is now saturated with those ex servicemen who were tradesmen like cooks, barbers, and the like and thus have hardly any professional expertise in security matters. To add to the worries of the of commanders of these Service installations there is a regular thrust towards reduction even in this rather thin security provided to the depots and work shops spread over thousands of acres of land and holding costly stores and equipments. It is understood that, 'in order to further cut the 'already small' tail there is a thought process to reduce the combat man power and have the 'base work shops' etc. run exclusively by a civilian staff. Generally the industrial and non-industrial workers are good. However, the heavy thrust of trade unionism, caste based intakes where professional efficiency is seconded to other considerations, impact of politicians, and what is 'wrongly' described as "work culture" results in productivity being severely restricted. There are depots where the industrial workers contribute hardly 3 to 4 hours of productivity in a day. The management is held to ransom if any effort to create a disciplined environment is made. (When author was the commandant of a 'highly politicised' depot all kinds of VIPs used to visit the station for a holy dip. During an informal conversation with one such visiting dignitary, it was suggested that if instead of 3000 labourers the depot was given 600 hundred combatants a better job could done. The depot could also have much higher level of assurance and productivity. The visiting VIP observed that those workers and their families put together make a decent 'vote bank'. "Of what use will the combatants be to a self respecting local politician when he had no way of getting them registered as voters in his constituency"? He pontificated). The underlying implication was that in so far as the higher management of national defence was concerned, higher productivity was a secondary consideration. The issue of reducing combatant manpower from Service installations needs to be examined taking into consideration such like factors as well.

Secondly, there is a need to appreciate that if we create overly civilianised organisation in Service units the pool of trained and experienced military work-force in the field of equipment maintenance, ammunition repair and handling, storage and ware-housing operation, supply management will get smaller. This could create problems when skilled combat manpower is needed either for new raisings or when strength in logistics installations needs building up for ensuring material-readiness of the field formations. We do not seem to appreciate that it takes years and years of dedicated effort for an individual to become a competent electronics engineer or armament artificer. We cannot pick such "equipment experts" off the streets if urgency suddenly arises. A decent complement of combat supporting element is a necessity that should not be ignored while deciding on the size of our military tail. With ever increasing demand for enhanced support to the front line soldier and with ever decreasing manpower and financial resources 'Services' are finding it more and more difficult to measure up. Secondly, how the Administrative Services are to be manned is increasingly having a significant effect upon the development of the officer structure of the armed forces. Service installations face tougher challenges than before. The problem is two folds; first, the Services demand a wide and diverse range of skills that are not normally required by the fighting men. Second, the situation of resource limitations and largely uncertain troop commitment in these days of prolonged 'IS' duties demand a very high level of innovation and training in various facets of resource management. These skills take time to be imbibed and make one grow to be a competent material manager or technician.

Impact of High Technology

The most important fundamental is that the industrialisation of war has made the organisation of logistics element in support of defence forces so large that its sheer size creates special problems. As the technology advances it demands more and more logistics support and greater logistics lead-time while at the same time creating the possibility of stockpiling obsolescence. In our case the situation is more acute because decision taking for acquiring the latest in the field is so time consuming that very often we continue to keep living with antiquities in an atmosphere of uncertainty thereby causing a stretch on our support Services. Finding spare parts for those equipments which had been declared as obsolescent and then again brought on active list because either there was no money for replacements, or sellers were not forthcoming, has made life extremely difficult for AOC &EME. We should appreciate the fact that the research, development and procurement costs of new equipment are often less than those of operating and servicing the systems which are over their useful lives. This is particularly true of more sophisticated systems that require a large number of skilled maintenance personnel and holding of spares and subsystems which are expensive with a potential for becoming useless in due course of time. A leading defence analyst expressed the decision-maker's problem with characteristic style in the following way: 'A long-range military problem is comparable to the problem of a racing stable owner who wants to win a race. To be run many years hence, on a track not yet built, between horses not yet born. To make matters worse the possibility exists that when the race is finally run the rules may have been changed the track length altered, and the horses replaced by greyhounds' implying that the horses may go out of service.[1] We have that kind of dilemma today, because we have been acquiring equipment over the years from any and every acceptable source. This has created a situation where we need experts in too diverse a field to be practically and economically viable. Therefore, any reduction in our support services must take into account that the actions taken today may have tremendous impact on our support system ten years down the line. It may consume much more money and effort at that stage to create the infrastructure that we may dismantle today.

Another cause for concern is that commanders in their zeal to reduce administrative tail are generally propagating large-scale reduction in the combatant manpower in the service units without giving adequate thought to the adverse effects such a policy can have on the fighting capability of the front line soldier. Today we may reduce the size of our 'tail', but we must realise that replacing skilled individuals with persons of like skills when a requirement comes up will be difficult. When a situation arises where we have to have many more technicians for our combat formations the army will be faced with the problem of training thousands of personnel with entry -level technical skills. We will also have to provide additional training in lieu of 'skill progression' normally acquired by 'on-the-job' experience over a considerable length of time. Even thereafter the technician is not likely to be a 'seasoned' expert. Material support to the military is procured at considerable expense. These resources are in the form of material, equipments, services and a very wide range of national assets. These materials are to be kept battle-worthy at all times. Guns, tanks, aircrafts, radio equipment are all machines and these machines are not like men. Exhortation, personal examples, leadership etc have no effect on them. They need to be maintained, sustained and looked after at all times. The problems with maintenance management will only worsen with induction of higher technology because more intricate the technology, greater is the need for better class and number of technicians. Therefore, as the Military becomes a more high-tech organisation relying more on complex weapon systems than individual soldiers to destroy the enemy it must focus on development of technical manpower and technicians. To maintain and sustain the highest level of functional efficiency more and more men skilled in their assigned technical functions are required. Increased technological developments and our need to modernise our Army have created a situation where there is urgency for rapid large-scale industrial and training operations to be ordered up on short notice. Instead of doing that we are finding ways of reducing our tail without giving adequate thought to the increasing requirement of technical manpower. Thus, in the absence of a more realistic assessment of our requirement we are faced with a real dilemma of a need for creating a huge logistic organisation that is both flexible and highly responsive to sudden operational urgency and yet provide a shortened tail.

Right Sizing - How to go about it?

In all the modern armies of the world the fact that sustaining a combat soldier on front-line is not a subject that could afford adhoc-ism has come to be accepted. In some cases this has happened as "lessons learnt" from some campaign or the other; or as a result of serious 'soul-searching'. The repercussions have been discussed above. We can achieve some reduction in our base depots etc. However, we should not take up the exercise without a detailed examination of its likely after effects. Certain basic questions need to be tackled first if we really want to have a lean and mean, and yet effective material support system. Broadly these questions are:

(a) In the case of mobile formations, for accomplishment of their mission, do we require the entire range of their 'entitled' stores to be carried into a battle? The subject has been thrashed in various exercises. After identification of specific material requirement for various phases of an operation, priorities, load-tables, and transportation allocation etc. have been made in the form of 'standard operating procedures' (S O Ps). The fear of an unforeseen situation still exists at all levels. Consequently, a lot of hardware and iron-mongery gets prioritised. This leads to a situation where certain items that do not materially add to operational efficiency of a combat unit get added to the list. This adds to the tail complement at all levels.

During the course of technical inspections of certain equipment intensive and operationally committed units the author has seen them holding on to a very wide range of stores which they can never need before, during or after an actual operation. Their "equipment tables" authorise them these items for normal maintenance during peacetime and in their "peace" locations. Since the equipment tables have specified certain scales (or with misplaced faith in the quartermaster's judgement given the authorisation "as required') the unit quartermasters keep demanding unreasonable quantities regularly. There are units holding as much as 30000 kgs of 'boot-nails' when the hob-nailed boots have virtually gone out of service. One could come across an artillery regiment (now equipped with a different weapon system) still holding spares for the 25 pdr gun on the premise 'one never knows when a particular need may arise'. Such situations do create a certain amount of strain on the Support Services and those which are integral to a field formation have to find ways and means to measure up to these idiosyncracies. Once in a while the unit commander or even the brigade commander must walk through the stores so as to see what all amongst the "held inventory" will they actually need when the unit gets deployed in operations. Otherwise, it will continue to happen that 'street smart' quartermasters will keep demanding and hoarding items which could only be described as 'cosmetic' without any specific operational utility. Remember that an unneeded material or resources clog the provision and distribution of needed resources. This also means additional workload at all levels of material support demanding more men, material, and transport.

There is a definite need for an analysis in order to identify absolute essentials separately for exercises - where social graces tend to play a more important role-and what will definitely be required in an actual operation. This will dictate the levels of reserves and priorities of maintenance of a field-formation when engaged. It will also tell us the kind of support service needed during a lull in the situation. We only have to resource front-line units to sustain combat with critical material. Rear area logistics can sustain the not so important material that ensures adequate health and comfort in not-so-critical locations. In the forward areas supply stocks should be confined only to fast moving military essentials, supplemented by a small supply of those slow moving items which are critical to maintenance of the combat readiness. If these stocks are backed up by a responsive, reliable, fast transportation system, efficient and effective support can be furnished at much less than the cost of attempting to provide all the items authorised in the equipment table of a combat unit. Remember that stripping the supply system to its non-essentials attacks the logistic snowball at its roots and ensures toughness and resiliency instead of fat and sluggishness in the logistic support system. It will be important to stipulate that in the matter of reducing the number of items both in terms of depth and range the technical officers have a meaningful role to play. They have to identify critical equipment, weapon systems, and repair parts needed to sustain tactical combat units and include only those items in contingency planning.

The logistics officers and technicians must ensure that the method of supply and demand control takes into account the difference between usage in peace and usage in war. In this we also need to revise our ongoing philosophy where 'provision action' for first five years is based on manufacturer recommended scales and thereafter, on actual usage. Since the actual usage, particularly in the case of imported equipment, is restricted in order to prolong its 'life-span' no realistic data- base gets built and the provision action tends to get faulty leading to all kinds of problems at various levels of the material support. We must realise that a system based on only peacetime and restricted usage with little or no provision for quick wartime prediction, supplement, and adjustment will give a fictitious sense of economy, and could be a big let down in an operational situation. This aspect assumes special significance when the 'Army Standing Establishment

Committee- ASEC' during the organisational audit of Service units and depots tends to take these deceptive workloads as "operational actual" and effect reductions in the manpower of the service units.

(b) Is there a need to re-examine integral servicing elements within Infantry/Artillery/ Armoured corps units? For whatever reason as and when a new artillery or armoured regiment is raised we tend to also raise a dedicated workshop with its full complement of men and material along side. This may give an enhanced confidence level to the unit commander and may also provide adequate experience to the 'tradesmen' in order to make them more useful in operations. But in an environment where most of the equipment tends to be 'moth-balled' this appears to be a wasteful and expansionist approach of the Support Services. More so, when during peace time generally the units holding a specific weapon system are located in close vicinity of each other. In order to keep a control on the size of the tail it may be a better idea to have only a 'core-group' of technicians specifically trained for the main equipment available in peacetime. The required strength of other categories can be built up if an operational situation arises. One question that may come up is that of the training of these personnel who only come in when a need has already arisen. What we need to examine is the fact that, in an intense engagement is there a requirement of carrying out 'field-repair'? If this is unavoidable then can we ensure serviceability of the damaged equipment with least effort and time during the course of an active operation?

(c) Will the dual trade capability help in reducing tail? The Service complement is usually fairly in the rear even in an active hostility. Therefore, in any operation considerable time elapses before the service element catches up with the main force. During this phase the forces are open to the enemy counter-attack, jitters and things like that. If their equipment is not restored to its serviceability at once there could be problems. To overcome this kind of situation all over the world the concept of 'dual-trade' is catching up. Various operations have shown the need for the combat-arms soldiers also to be able to carry out certain urgently needed repairs to their equipment on their own. Gunners must know how to repair their artillery pieces. Drivers must become mechanics. (In USA, Marine Corps philosophy is that every marine is a rifleman first, then a cook, a driver, storekeeper, fitter and so on). The Americans also have `dual' trade soldiering - be it rifleman or storekeeper. This ensures higher survivability rate for the rifleman till his logistics support fetches up and the storekeeper is defensible in case the logistics installation is attacked in any manner. If we train a driver to repair his vehicle by expedient means, we reinforce his need and the capability to take care of himself. Instead of looking for someone else for help, a driver with such training will try to handle his problem by himself. To take the idea a bit further, it is recommended that our military drivers should be sent with the civilian "truck drivers" when they go on long hauls say, from Delhi to Leh etc. Our boys can learn a lot about restoring serviceability of vehicles en-route when a radiator pipe leaks, or gear system breaks down. One sees enough of ingenuity, innovativeness and a spirit of mutual help in the 'truck driving community'. Similarly, Radio operators must know how to make one good transmitter from two damaged ones. In today's environment when 'modular' concept in respect of a very wide range of equipment has come up it should be possible for an infantry-man or an armoured corps sawar to restore the serviceability of his weapon or equipment without waiting for the arrival of his support-services. We should train our soldiers to be more self-reliant. However, the artillery, armour and air defence weapons today mean more wounded than killed demanding integral medical aid of a different kind. The medical services thus require somewhat different integration. Similar is the case with the spares and repair efforts needing heavier machines to carry out their tasks quickly and efficiently.

(d) Is there a need to introduce palletisation in a big way? Or could we go in for the 'Container System' so that the urgent requirements of the front-line soldier can be met in a cost-effective manner? The demountable racks now in use in most of the western armies can be loaded or unloaded in just a few minutes. Thus the productivity of the transport companies can be increased by as much as 60 percent. Also, instead of concentrated frenetic activity in the dark to load a conventional vehicle cargo convoy, the flat racks can be loaded during the day prior to the arrival of the transport, thereby doing more with less and in a much more operationally safe and effective manner. With the introduction of MHE and containerisation, a real reduction in the size of the tail can be achieved without affecting the efficiency in any manner. Therefore, we need to examine the aspect that improved 'material-handling' techniques reduce the administrative details and yet make them more effective, particularly in the plains and the deserts. In the mountains, depending on specific needs and the terrain, appropriate MHE can be introduced.

One way to reduce the tail and yet make it more effective will be to introduce equipment for handling thr neavier stores like 'palletised ammunition'. It must be understood that the firepower and therefore, ammunition consumption of a modern division is some four times its Second World War equivalent. Availability of pallets where 4 to 6 times more ammunition can be handled at a time reduces the need for support element in the forward areas. Management of material, repair and maintenance facility, medical covers, supply of rations etc., could certainly do with induction of technology in all these spheres. We must critically analyse which Service we should continue to provide, which ones should be contracted or consolidated and how to better afford and support that we retain. One would always agree that with urbanisation, better road communications, and improved civil infrastructure there is no need for something like "military farms". Or with the improvement in our road network in remote areas and consequent reduction in usage of mules do we need the "Remount and Veterinary corps". The diary farms could be retained or shifted to those remote locations where the facilities aren't as yet adequate and be managed under the aegis of "Q" staff. Similarly the animals could be handled in their areas of employment. For that matter with the educational level of the recruits having considerably improved why do we need an "Army Education Corps"?

(e) Should we make use of civil infrastructure in our cantonments? Today, almost 2/3rd of the MES budget are expended in salary of the staff. Similarly, a large workforce is employed for maintaining our cantonments. The staff hired for the purpose is usually erratic and absence-prone in its behaviour because of intense trade union and political patronage. There is merit in contracting out these services to those who can provide much cheaper and more efficient service to the troops and their families in a reliable and efficient manner. We must seize the initiative to operate our logistics installations, military garrisons and stations in a smart, business-like manner in order to provide our soldiers and their families, civilians (in defence employ) and others the best possible living, working and training environment in our cantonments, and other military stations.

Focus on Common Training

A worrying factor is that, even though officers & men from the Services, have a different role and tasks to perform their training & grooming some how does not seem to make them survivable in future war. Unfortunately, the logisticians think that what they do in peacetime i.e. honing their technical skills is what they will need in war. They tend to neglect the fact that during an actual war enemy's main thrust is likely to be to incapacitate the support capability of the front-line forces and thus take control of the situation. Ammunition dumps and petroleum depots are ideal targets for the enemy's air. If the ammunition and fuel are not forthcoming then there is no reaction capability. Shear size of these installations makes it virtually impossible to conceal them. The enemy is bound to know these logistics installations. Since the fighting arms are likely to have their hands full elsewhere these Service units will have to defend themselves; and may be, even have to beat back well-planned attacks both on ground and air. Equally important is for the forward troops to have the capability of sustaining themselves till the support services fetch up. There could be any number of hold ups in so far as movement of the Workshop machinery lorries or the Ordnance complement, medical facilities, Engineers for developing roads and tracks or the Signal troops for laying cable etc, is concerned. In the meanwhile, there could be radio-sets needing urgent repairs or vehicles needing spares due to the enemy action. This is a dangerous phase and the troops have to be capable of beating back a counter-attack. Similarly, the Services also have to understand the need for them to be on their own for a long while during an operation. This means that the combat support soldiers need to know how to lay mines and booby traps and to properly employ their organic weapons. They should be able to call for and adjust artillery fire, mark their positions for air support and provide for passive air-defence so that the enemy's assaults can be beaten back. These are training tasks that will help to ensure that service-personnel become " survivable" so that they can perform their real mission-support for the combat soldier. For this we need to have training programmes where both sides of the coin could look into each other's capabilities and limitations. Unless there is a unity of purpose in the training programmes and functioning this kind of efficiency cannot be achieved. Therefore, what we need is more of cross training so that combat-arms can handle immediate logistical problems without having to look over the shoulder, and support services do not have to await arrival of 'protection parties' and other combat experts before they set shop in an operationally active area.

Surprisingly, the prevailing philosophy seems to be to grow leaders and leaders only rather than adapting to the cultural change to also grow technical managers and technicians. If we want to ensure a high level of logistic-support then we have to accept the fact that technical training of the same magnitude as the situational training given to the combat arm leaders is what is needed in the technical arena. For a professional army this is important to have material managers and technocrats who are equally efficient and proficient soldiers. That is to say that there is a need to organically blend the combat and support services and make them a unified and effective war machine. Then only we can affect any worthwhile changes in our 'tail to teeth ratio' without any adverse effect on functional efficiency of the combat unit. The logistician and the tactician need to appreciate the requirements and capabilities of each other's organisation and equipment. Such an understanding will help in identifying tactical plans that cannot be supported logistically. It will force both to prepare plans that make most efficient use of our combat, combat support and service units. What we need is more cross training. Today, we think and train in terms of "tooth" and in terms of "tail" and not as an integrated team. Training programmes must provide a common focus. Individual deficiencies must be identified and corrected and the art of war must be practiced as a syncronised force. Combat arms schools need to place greater emphasis on transportation, maintenance and effects they have on tactics and tactical plans. Similarly, "Service training establishments" need greater thrust on tactics and proficiency in usage of various weapons. It is time we accept the fact that 'logistic soldiers' who are undisciplined and untrained in selected soldier skills will probably not survive enemy action to employ their technical skills. (I have never understood the rationale behind different scales of training ammunition for the combat and service soldier. I have also not been able to fathom any reason behind having only a token presence of service officers on 'All Arms courses'. As a matter of fact service officers are not even considered for detailment on courses either at the Armoured Corps or Artillery training courses. Logically, since these two demand maximum support during operations joint training with ammunition experts of Ordnance, " vehicle experts of EME and petroleum specialists of ASC could clear a large number of basic and fundamental doubts. For the combat soldier reverse is equally true. He must be trained to perform selected logistical skills to prepare for the time when normal means of re supply are interrupted. If we do not train to fight as integrated warriors we might not win. (According to Lt. Gen. John R Calvin vii Corps US Army " Logistic units are going to be doing more fighting and combat units more self help logistics than ever before. As never before the tooth and the tail dragon must fight and survive as a balanced and integrated warrior".)

One thing must be clear to all of us. That is, if the Army is to be successful in 'right-sizing', then we have to grant that logistics is a team effort that must include all logisticians and operational commanders at all levels. There is a need to bridge the communication gap between the tacticians and logisticians. We have to produce military leaders who understand the interdependence of tactics and logistics. How so ever hard we debate on the 'tail to teeth ratio' the fact of the matter is that manpower and resources are essential for meeting the support requirements of combat forces. A modern army cannot be all "teeth and no tail". A modern combat force is one whole, combat troops and logistics. In order to prepare for war we must provide for and define the practical tasks of equipping, boarding and lodging, providing medical care, maintaining and sustaining the forces under all kinds of situations the armed forces are expected to operate in. We must assign these responsibilities to specific organisations and individuals that have been trained and groomed to perform these specific tasks. For logistics to be responsive interplay of all the factors must be understood by command at whatever level it is being exercised.

Very often one hears of some 'hot shot' commanders arguing that most of the support services are irrelevant because these commanders do not get the level of support which their tactical brilliance demands. There was a famous corps commander who saw his Service Heads (other than the medical representative, of course) only on arrival in his formation or when they got posted out. In so far as he was concerned they were only "decoration pieces" while attending his conferences. They were neither expected to have any opinion on matters military, nor were they expected to express any. Even where the subject concerned them or was relevant in the realm of material support they were to put up through his "general staff" channel. It is not difficult to see that, at times, why the logisticians do not provide adequate support is easily attributable to the fact that they hardly communicate with the supported commander. Frankly, they do not speak the same language, and, therefore, the understanding of the requirements, responsibilities and results is often different. Consequently, it is not uncommon for a tactical plan to be developed without an in-depth analysis of the logistics involved. Something known as "Adm check" though carried out does not often educate the commander with regard to feasibility of his operational plan because those who have limited exposure to specific logistics tasks carry out this check.

The requirement is that the staff, assisted by the specialists must sort and analyse the information before presenting it to the decision-maker. This process takes valuable time, slowing the analysis and presentation process. If the services are not involved then there are chances that certain critical information will be overlooked or missed in the frenzy of processing large amounts of information. All this means is that commanders at different levels have to ensure adequate importance to the efficiency and competence of their service elements. Then only logistically supportable plans can be made and put into operation. Unless the logistician knows, practises and perfects all that is essential for him to be useful to the combat-soldier he may not be able to perform his real mission-support for the on going operation. For this to happen the Military must change its traditional ways of thinking. Very few commanders appreciate that today there is a greater need for the tactical commander to understand and accept the limitations and capabilities of the Support Services. In the absence of such awareness and in the absence of active participation at all levels of the professionals trained and hired to provide expert advice in their respective fields, realistic plans in respect of logistics can hardly ever be made and executed. We tend to play down the role of the logisticians, both during peace and war. We have the infantry as the 'queen of the battle', armour claims itself to be the 'king of the battle', and then there are any number of upstarts with a variety of claims towards exotic titles in so far as the pecking order is concerned. Regrettably, we tend to ignore the fact that these kings and queens are meaningless if they are not provided with wherewithal to live up to their designated roles and rather imaginatively seductive titles.

From the foregoing one thing must be clear to all of us. That is if the Army is to be successful in 'right-sizing', then we have to accept the fact that logistics is a team effort that must include all logisticians at all levels. There is a need to bridge the communication gap between the tacticians and logisticians and to produce military leaders who understand the interdependence of tactics and logistics. More importantly, we physically hold inventories worth hundreds of crores of rupees in various echelons. These stores and resources, procured at considerable expense to the state, need to be looked after and preserved in a manner that retains their serviceability for a prolonged length of time, and that means a greater level of officer level involvement. And yet, we have accepted a situation where the officers commissioned into ASC, AOC, EME etc., are posted to infantry units for 2/3 years. May be doing an infantry attachment for a years or so is good for a Service officer because it enables him to understand the problems of the supported troops 'first hand". This experience will stand him in good stead when he joins his parent service. But any period beyond one-year delays his professional grooming by that much length of time besides accentuating shortages in the Service units leading to numerous problems discussed earlier. We need to understand that the number of officers and technicians required in the support Services is large in relation to those needed in front line units.

It is not my case that we continue to retain uneconomical and man power intensive support where modern technology could give us more useful service at reduced expense in terms of men and material. The subject needs a total and pragmatic approach so that instead of these 'hard scales' and 'performance degrading' cuts we could have a scientifically evaluated support set up. All that we need to grasp is the fact that there really cannot be a tiger without a tail and the conduct and support of war now includes almost every activity and organisation in a nation.

Notes and References

1. *System Analysis and Policy Planning: Applications in Defence*, by E. S. Quade and W. I. Boucher, p. 418.

9

Present Support System

Traditionally, the system of logistics support prescribes base depots which are well into the rear, then there are regional depots that cater for a corps zone and units integral to a division. The current system such as Base Areas, Army Areas and the Corps Maintenance Areas etc. are based on the British experience of World War II with some peripheral changes brought about during the last four decades. In our endeavour to place faith in tried and tested methods we tend to overlook the fact that those times were different. The means of communication were restricted, the country hardly had any industry and the British having rather uncomfortable memories of 1857 were not particularly keen on encouraging any project that could eventually adversely affect the empire. War-like materials were therefore, almost always brought from the sources the rulers had their faith in. Except for the fact that during the 2nd World War when some kind of defence-oriented industry was, perforce, established the system was entirely based on environmental factors existing at the time. This led to the concept of "Base System" etc. But the situation has altered considerably ever since, and the methodology with regard to logistics supply has altered considerably since World War 2 and the procedures are still being evolved. The principal changes have been in the system of bringing forward ammunition and supplies. The traditional methods have been the use of echelons; where loads are brought forward from base to the regional depots by the third or fourth line transport. These are then off-loaded and re-loaded and carried forward to the "corps locations" in another transport and so on up to the forward dumps. This system is wasteful and outmoded in terms of time and transport. Today, it would be necessary to carry to the forward most location items like ammunition, food and fuel with minimum transshipment en-route. At best there should be a requirement to change and relieve the drivers only at pre-decided points. Effective logistics is military power in motion. Logistics must be geared to keep the tactical units mobile and effective.

Our old practice of holding stocks at the unit, direct support, and intermediate levels are characterised by vertical requisitioning and a myopic focus on each supply node. Little, if any, consideration has been given to the expansion, mobility enhancement and technological changes that are required for a well equipped army in any given situation in so far as logistics support is concerned. The need for a more responsive and vibrant system has always been felt. Consequently, there have been some experiments in the right direction. No tangible gains, however, seem to have surfaced; it appears, primarily, due to our reluctance to change from what could be described as a time - tested system. Efforts to try out certain modifications in the existing system have also been made wherein repair, medical covers etc., have been dovetailed into locally available resources. But these invariably have lacked in measuring up. Principally, because of the usual reluctance to change and the fact that piecemeal changes never can give a total picture to commanders at various levels with regard to the efficacy of something radically different from what they have grown up with.

With inventory automation coming in, and when all logistics units get interconnected and multi-nodal there will be a need to think big. We will have to see (particularly in the case of troops engaged in 'internal security duties' and low intensity conflicts) that a customer's requirement can be met by a horizontal referral from a "Direct Support Unit" (DSU) that is not in its normal support chain. A unit shouldn't have to turn to an integral DSU(unless it is convenient to do so) to meet the demand- be it repair cover, supply of rations, FOL, or Ordnance stores, unless it is convenient to do so. Certain problems, like how is 'demand satisfaction' to be calculated will come up? Should it be based on the concept of a single node, or multiple, horizontal nodes? How should decisions on the depth and range of the prescribed PETs and WETs be made in a multi-nodal network of connected DSUs? At what level should these decisions be made? How should 'excess' be measured? What changes in the transportation system are necessary in a horizontal distribution system? And most importantly, how do we dispel a commander's fears that his critical assets may be diverted to other users? And finally how can peace-time "asset-vision" technology be used in war time and crisis situations?

Some of the questions that will come up when we decide to change, what has been described as a 'time tested', system have been listed. While the author was responsible for giving 'Ordnance cover' to a corps engaged in counter insurgency all these problems had come up. Particularly when induction/ de induction frequency got to be rather unpredictable and the units were in no position to follow the existing procedures where they are required to cancel demands in the previous locations and place afresh on arrival. It was found that the 'so called' time tested procedures and systems were impractical in the given situation and horizontal affiliations were most effective in ensuring timely logistic support. There were certain commanders who didn't like the idea of their stores being made available to "outsiders" as they apprehended reduction in their operational readiness, particularly where 'short supply' items were concerned. In such a situation the higher commander should dispel such fears and the 'Support Services' should 'guarantee' that the items thus consumed are made available at the 'Regional' depots, so that most economical use of inventory in a theater is ensured, while allaying apprehensions of the immediate commander. There are chances that the stores demanded earlier in the previous location may also materialise, thus leading to surpluses with the unit quartermaster. Services representatives at the Divisional levels will have to keep their eyes and ears open to ensure that this does not happen. Such surpluses need to be immediately merged in the DSU stocks.

What ever be the size and relative importance of the territory in which the operations are likely to be conducted, all supply machinery ultimately must stem from the main administrative bases. Because of diversity of various regions, peculiar means of communication in any region and limited availability of stores coupled with possibility of short intense wars there is a need to examine a pattern wherein a commander does not have to restrict his plans because of logistical constraints. Too often it has happened in the world's history that with the logistical support having broken-down the enemy has had a chance to recover, repair and reconstitute his forces. The seesaw battles between the Germans and the British Eighth Army in the African desert during World War 2 is just one example showing that " logistics always affect the tempo (and) the rhythm of battle."

Requisites of an Efficient System

- The organisation of the rear must reflect the character of the war and the nature of fighting. There cannot be a common setting for all kinds of situations. However, it would be appropriate to list some fundamentals that define an efficient system. These are:-
  - The higher commander is responsible for supplying lower formations, units and sub-units. It is not unusual for commanders to wash their hands of the whole affair once they convey their requirements to the service heads. In so far as they are concerned thereafter it is not their problem. Never having found time and patience to get a grasp of the issues involved these commanders do no good to either their troops or to their mission.

- All available transport must be used. There will always be a problem of finding appropriate means for conveying stores from one location to another. This will be particularly so where transshipment is involved. Since the terrain conditions where impending operations are planned will be known it will be sensible to have the packing of stores commensurate with the final means. That is, if eventually the stores have to be man-packed it will speed up the things if the bulk packing allows for a planned breaking of the main load into suitable loads. Equipment repair assets must be deployed in those areas where greatest number of repairable equipment and vehicles are either held or envisaged.
- The medical services must deploy as near as possible to the areas of largest number of casualties. Foraging for local supplies, and determination of local medical resources should be undertaken in peace time itself.
- Mutual logistics assistance, and horizontal logistics support should become a basic guiding principle for all concerned.

Ammunition supply will always assume critical proportions. Therefore, decide before hand the demanding procedure. That is, will it be automatic or on demand or a mix of the two. Allocation of transport and material handling equipment at the respective levels of supply at the formation and unit levels must be made in advance. Alternative system for urgent resupply if the on going system breaks down must be identified in advance. Ammunition packages that are designed for air-dropping are so marked. It will be appropriate to keep these segregated from other types of packing. These can be held in hand for a likely requirement of dropping ammunition by air. Replenishment of missiles involves a large number of activities at the issuing depot and the user unit level. These should be well-rehearsed before any engagement.

Structure, organisation and manpower for logistics units be decided before hand. In inhospitable areas locals are better equipped to carry heavy loads. They be employed as labour. In this ex-servicemen belonging to that particular area can play a useful role in locating and bringing sturdy and hardworking locals forward for loading and unloading purposes.

Logistics Functions

Functions of logistics can be broken up into five main groups, ie Supply, Transportation and movement; Maintenance and repair; Medical; and the smaller functions like post, catering and labour.

The army of today, not only has extremely sophisticated and costly equipment but its operations are likely to be vastly different from what they have been in the past. Technological developments dictate a battle field scenario for the future where dispersion and decentralisation of command and control will predominate, space for manoeuver will be extensive, and the requirements for concentration of the force will be dictated by speed in the shortest possible time. In such an environment, logistics support will require to be flexible enough to cater for achieving these quick concentrations and dispersions to enable commanders to fight effectively. This, inter-alia, would imply commensurate mobility for logistics backing for an operation which could be of a reasonable (but short and intense) duration. The need today is not only to outgun the opposition but also to out think vis-a-vis logistics support. The subject has generated enough debate, interest and perhaps, concern amongst commanders at various levels. It is, therefore, necessary to discuss the logistics support required and how this could possibly meet the conflicting needs of economy and a higher level of assurance.

At first sight it may appear logical that prepositioning of all resources either in or within land-communication of the potential zone would provide complete answer to the problem. But any such solution presupposes infallible appreciation of all areas in which the battles will be fought and their likely duration. Not to be forgotten is the fact that there is the other side also who will not necessarily fall into the predicted tactical appreciation of an impending operation. Complete reliance on "prepositioned resources" will automatically deny any flexibility in the operational exploitation of the situations as they arise because the other side is not likely to conform and there are far too many unpredictable situations that will invariably arise in any operation. To have enemy play to your tune is obviously impossible. Dependence only on prepositioned support would be virtually a disastrous basis on which to plan your operations.

Fundamental to the successful conduct of a military operation is that a commander must have arms and ammunition required for his task in hand. He must have resources for maintenance of his equipment, vehicles and accessories, he must have food for his men and finally he must have means of delivering all these to the areas where he intends to fight his battle. His plans will include the likely duration of an impending campaign, and based on a predicted consumption pattern of fuel, rations and ammunition, he will make his requisitions which, after they have been examined at an appropriate level may either be met in full or moderated in consonance with the higher commander's over all plan. Availability will also govern what is requisitioned and what is made available. Of course, the commander will have to have enough elasticity in his overall capability to switch axis if and when such a need arises. The task, to put it mildly, is monumental because to cater for all kinds of eventualities will be grossly uneconomical and can in no case be acceptable. It can be said that some system of administrative replenishment, other than total pre-stocking, is essential, and a certain amount of stockpiling will, undoubtedly, be necessary.

This means that the Army Logistics Organisation must improve operations to the point that the primary system constraint is the market. This would mean that the organisational system is fully effective and able to meet all demands placed on it. In other words, the system as a whole would have excess capacity. All customer demands would be met completely and on time by the organisation, and the excess could translate into a wartime surge capability. We know that in our case the situation is far from ideal and the limitations of resources will always place greater demand on appropriate and economical use of all the resources. We need to have a system that meets all the operational requirements of combat soldier and also provides a high level of assurance to the commander allowing him to exploit his success, be flexible in his tactical plans, and yet not tied him down with heavy baggage that he cannot lug around when a battle is on. Various combinations have been tried in other armies to meet logistics requirements of combat troops. A system, which is required to meet material needs of the troops, is generally designed on the following lines.

1. **Base Support System** - Military strategic depots and national industrial bases form the core of the logistic support. They are all linked by the 'lines of communications'. 'Up' the lines of communications flow reinforcements and supplies. 'Down' them flow the causalities and empty supply vehicles. Thus, the lines of communications see a continuous two-way traffic along which the men and material move.
2. **Intermediate Support System** - This support is provided in the form of 'Regional Depots', workshops, hospitals and certain other ancillaries. These units are sited out of range of the 'depth fire' artillery of the enemy. Assets are widely dispersed and depending on the role of supported formation the units could also be 'truck-mounted'. This provides adequate mobility to the

supported commander in fructifying his operational plans.

3. **Direct Support System**- We have the "Divisional" support units with their limited range of stores and prescribed man-power. These units are normally tailor-made for 'mountains, plains, deserts, and any specific role that has already been identified. These units do not have sufficient elasticity, and flexibility to undertake a "diversified role" or additional responsibility, certainly not for an indeterminate length of time. In some cases, due to operational or situational considerations, we do have detachments spread over a vast distance- an arrangement that spreads resources rather thin on the ground. In some cases due to terrain considerations and troops deployment that is of more or less of a permanent nature these detachments continue to perform for years on end. Innumerable problems regarding manpower, shortages, inadequacies of the system manifest themselves from time to time. No uniformly acceptable solution has been found so far. For a "Direct Support System" to be economical, effective and efficient we need to consider the following:
   - (a) Higher availability
   - (b) Shorter lead time
   - (c) Accurate MIS
   - (d) Should preferably be a 'single window system'
   - (e) Should have efficient thru' puts
   - (f) Regulations be flexible
   - (g) Needs to be capable of responding to 'operational needs'
   - (h) Soldier technicians or dual trade needs to be adopted as a norm

In so far as field formations are concerned there primarily are two systems of logistic support being practised in most of the developed countries. These can be put into two categories:

(a) **Pull System**: In this, combat supplies are sent to the front only on receipt of a request from the fighting units for their specified needs. Stores are usually held in an acceptable turn around time frame. Largely, spares, accessories, ammunition items, defence stores, additional rations and fuel form part of the items that can be on demand. First and second line stores are carried by the units themselves. It has following merits and demerits:

Merits of the System

- It exercises an economy in resources because only the needed stores are demanded and supplied.
- Affects a saving in combat manpower and transport utilisation as the collection parties are sent purely on 'need based' collection missions.
- Can reduce the tail element of the fighting troops where a commander can decide what all to carry along and what should be supplied to him on demand.
- Reduces chances of combat supplies falling into enemy hands because only a selected range of stores are carried in the first instance and the rest made available as the situation develops.

Demerits

- It does not provide flexibility to the commander. While making his initial judgement the commander could easily err in his selection of stores that is usually based on an assessment of a situation which is only likely to arise. It may not happen in the predicted manner, thus denying him the chance of exploiting a developing situation.
- Creates uncertainty because the gap between the combat troops and their logistics support could have widened due to operational situation, thus

putting the "time and space" complement out of schedule by consuming more than the acceptable and predetermined time frame.

- Likely enemy interference in the line of communication could also affect the supply chain.

(b) **Push System**: The system has been discussed at length earlier. In this, the planned quantity of supplies are sent to the forward troops at a regular interval. These quantities are based on the impending operations and likely daily expenditure.

Merits

- Provides flexibility to the commander as he is not uncertain of his material support.
- Ensured supplies and allows for timely planning.
- Facilitates in prepositioning the combat supply stocks. Since the details of incoming receipts are already known a commander can plan what to locate and where.
- Gives confidence to the troops as they need not look over their shoulders for supplies.
- Allows a combat commander to concentrate on actual operations.
- Reduces chances of a stock out at the bases because the requirements would already have been provisioned for.
- Impressment of civilian resources can be planned in advance. (This is a big plus because with a sufficiently advanced warning of military intentions the local civil administration can take appropriate action to ensure that the local population suffers minimum discomfort due to a military need).
- Facilitates positioning of rolling stock for movement of military stores.
- Reduces procedural intricacies as very limited surprises are given to the supplying agencies.

Demerits

- Overloading of the fighting troops is a serious problem due to advance stocking as the system would automatically be sending stores forward whether a requirement comes up or not.
- Not cost effective because stores are moved forward based only on a predictive requirement which may or may not actualise.
- Flexibility of the logistician is reduced because he cannot meet "suddenly arising" needs as the holdings are already committed.
- A change in the operational situation could have led to a change in destination while the stores are in transit. It may have become difficult to alter direction at that stage.

Applicability in Our Environment

In our case with such a diverse range of inventory and operational commitments in different regions and perpetual shortages it may not be advisable, cost effective, and practical to go in for any one system independently. In the case of mechanised forces, since they are likely to operate in deserts, semi deserts and those plain areas where communications are not fully developed it may be appropriate to adopt the "push system" initially for the strike formations and when the battle condition stabilises in the depth to go in for the "pull system". It is, however, imperative that logistic support units are provided with greater cross-country and matching mobility so as to ensure delivery at the desired time and place.

Similarly, in the case of formations deployed in the plains it might be economical to put some reliance on the civilian infrastructure for common user items, medical facilities, and rations etc. In so far as logistics support to the troops deployed in mountains are concerned, depending on the operational and designated role of the combat troops and nature of terrain ie is it prone to isolation because of the climatic conditions or any other reason will decide the system to be followed. In case the terrain is excessively difficult, airlift should also be planned and catered for. Counter insurgency operations and actions "in aid of civil authority" have their own peculiarities to be assigned any particular type of logistics support. These are neither sufficiently predictable, nor that 'well defined' for an accurate assessment to be made either for provisioning purposes or for undertaking supplies. All this means is that a mix of the two systems depending on the situation can give more efficient, cost effective and reliable material support to the troops. One has to ensure that a commander is provided with adequate flexibility, regular supplies, and is not required to worry about his logistics. For supporting up to an Infantry or Armoured Division a "Logistics Support Area" can be set up. Ideally a 'Logistics Support Area (LSA) for a division should have the following:

(a) **Established along the formation 'centre line'** - This will help in making 'services' available to a larger number of troops who will be able to reach the locations with minimal effort. The need for such a setting will be more predominant for 'desert operations' where the requirement for replenishment of ammunition and FOL will be far greater and mostly unpredictable. Heavier servicing machinery and ancillary lorries can be comfortably placed for the dependent troops to have urgent repairs undertaken quickly. Recovery of equipment damaged in action and evacuation of casualties to organic medical facilities will consume lesser time.

(b) **Adequate space to deploy approximately 400 vehicles at a time** - This includes the first and/or second line vehicles which will be coming for replenishment of ammunition or FOL. In the absence of space there could be cluttering up, leading to confusion, wastage of time and effort, and may even make the LSA a lucrative target for the enemy air. It is also to be kept in mind that in addition to making issues to the troops the Service units themselves will usually be replenished concurrently. In such a situation it might hasten things if it becomes feasible to divert the 'replenishing vehicles' (particularly those carrying ammunition and FOL) to the forward units instead of unloading these in the LSA.

(c) **To be on Hard Ground** - Often difficult to manage hard ground of desired specifications but effort has to be made to ensure that vagaries of the weather or any such reason does not lead to breakdown of the replenishment system. If need be, 'Engineer' effort can be requisitioned for the purpose. The in/out arrangements for vehicles be well defined and these be properly marked so that no confusion is caused.

(d) **Near a Water Point** - Administrative units themselves need a lot of water. In case it is not adequately available arrangements will have to be made both for 'potable' and water required for other purposes, particularly if bathing and washing facility is also set up in the LSA. It will be useful if wells, rivulets, other sources are identified and marked during peacetime itself so that when troops move for their operational deployments facilities are in place. In deserts water pipes could be laid under the arrangements of the concerned state governments. Local population should be encouraged to make use of these resources so that regular maintenance can be ensured at all times.

(e) **Adequate Cover**- It is not always possible to ensure but effort to locate service units in a manner which provides protection from enemy air is essential. Alternatively, it may become important to provide 'air defence' artillery for the purpose. In that case the service units should have their own capability to call for both air strike and direct fire on to encroaching enemy air-crafts.

(f) **Should be Defensible** - Service units are not likely to have spare combatant manpower for the defence of LSA, because most of the officers and men will be technicians who have specific functions to perform. However, it will also not be possible for the commander to spare fighting troops for the defence of LSA. Siting it in a manner that reduces the need for more troops will be a better option, even if certain other considerations have to be sacrificed for ensuring the defensibility.

(g) **Easy Communication** - It is not unusual for the Service units to be given a lower priority while establishing communications. A commander must ensure that he has an assured communication system with his logistics complement; otherwise chances are that wrong, inadequate and inaccurate issues could be made. That kind of situation could cause greater problems in the long run. It has been seen non-matching fuzes being issued with artillery ammunition because the communications were garbled and open to different interpretations. Where chances of a communication break down exist it might be a good policy to predetermine the nature, number and types of complementary fuzes, oils and gasses. These can be indicated to all concerned much in advance so that even if there is a break down in communications percentage of errors being made by the issuing agencies can be reduced. Service units must cater for the fact that continued and consistent inflow of munitions of war has to be ensured over a lengthy communication system from the base depots to the forward echelons.

In so far as the combat soldier is concerned he should always be aware of the system supporting him and his own role in getting supplies. Very often the situation demands immediate creation of some dump, air-drop location, or repair of the road damaged by the enemy. In such a situation the commander on the spot should ensure that necessary facilities are either available or are created for the technicians to carry out their task. In counter-insurgency operations and fast paced set piece battles it is not uncommon for the situation to change quickly and pre-decided locations become non-operative . The CMA or even the DMA may not even be aware of these developments. In a case like this an operation can either get stalled or the focus may change. To guard against such a situation it will be a better idea for the logistics installation commander to make himself aware of the latest on the front before the vehicle convoy sets off or pre-planned air drop undertaken. Very often in the absence of these details the stores have fallen into wrong hands. Vietcong guerillas virtually lived off the American's resources and American troops suffered a large number of casualties because inspite of a much more advanced technology and communications systems the guerillas could cause confusion in the enemy ranks and get away with the material meant. Contingency plans for such like situations must be made alongside the tactical plans. If possible these be rehearsed and refined before the enemy joins battle. The terrain, weather, availability of local resources all have an impact on logistics support to forward troops. And commanders will do well to keep this in mind when making their operational plans.

It is neither feasible nor desirable to design our logistics support system de novo. Where fixed assets are already existing on the ground these be suitably modified to conform to the chosen pattern. A mix of pull and push system of replenishments can give a greater level of assurance to the supported troops. The chain of replenishment as projected below should largely meet all kinds of situations. Nevertheless, it might be prudent to deal with a situation as it exists on the ground and logistics planned to conform to likely operational situations. Facilities and services created in any theater for defensive operations must have the capability to sustain offensive operations as well. Although, in the following chart the "chain of replenishment" shows 'Air Maintenance Cell' at command headquarters it might not be a bad idea to have these units affiliated to corps. Much larger spread of a corps' area of responsibility will dictate such an allocation.

CHAIN OF REPLENISHMENT

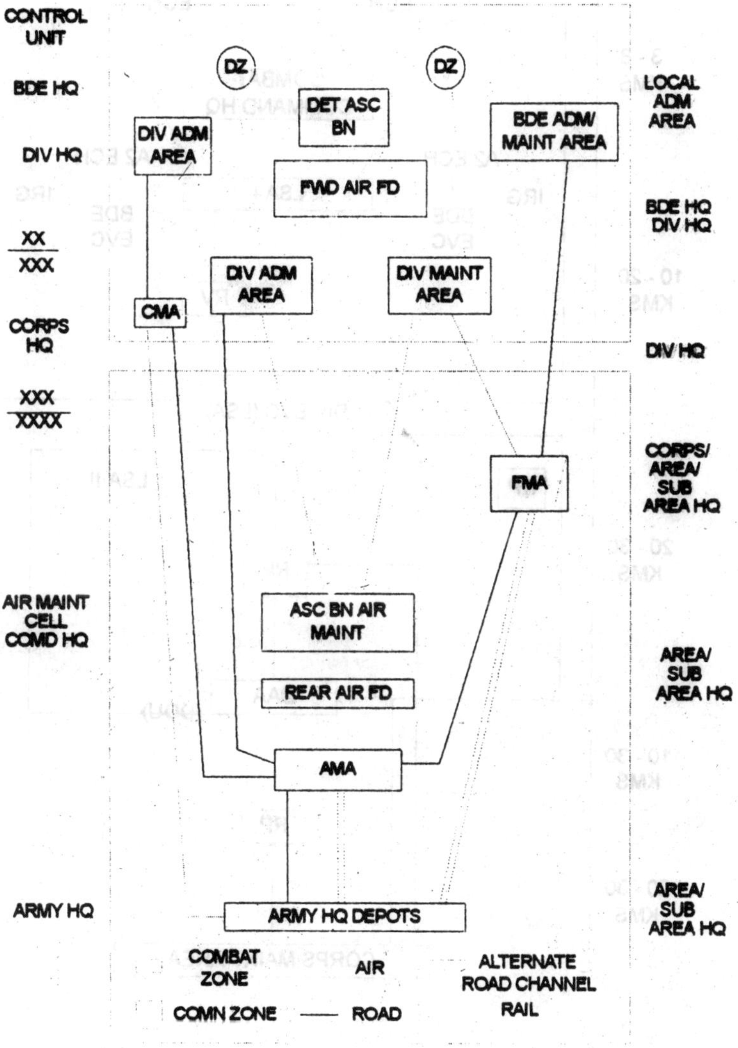

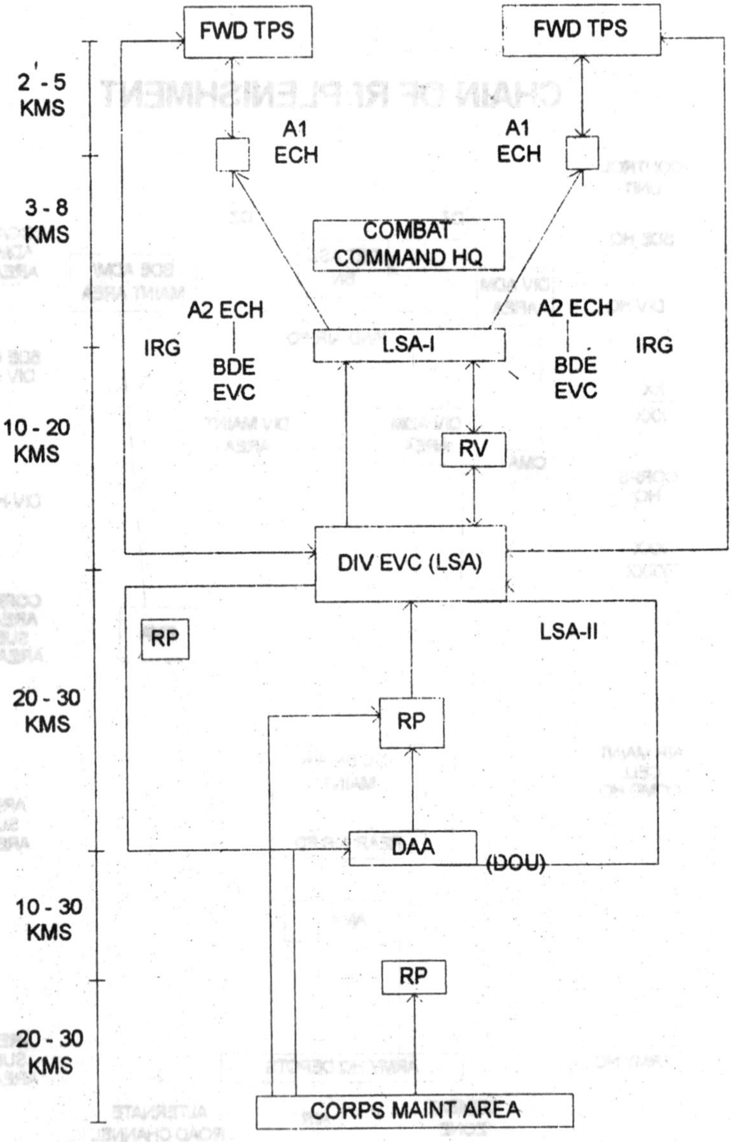

IRG Integral Replenishment Group is usually vehicle mounted and carries predicted essential replenishments.

10

Logistics - When Environment is Nuclear

In May 1998 both the tradition rivals- India and Pakistan- carried out their nuclear tests. Now that both the traditional rivals have come out in the open conventional war under nuclear threat will be different from our previous experiences. Nuclear weapons have surfaced in our region. Chances of any one of the adversaries being put in a dangerously provocative "use them or lose them" situation in a crisis cannot be ruled out. Now that the phenomenon has emerged in our backyard there is ample ground to examine effects of nuclear weapons on the troops, community at large, the military equipment and other resources, both military and civil; assuming that any one country is rash enough to use these ultimate destroyers. Logistics and material support assume an additional facet when there are chances of a nuclear exchange. This means a study of the future "battle-fields" because a nuclear weapon is a totally different ball game.

Problem is that any discussion relating to usage of nuclear weapons must take into account enormous uncertainties and imponderables that surround virtually every aspect of the matter. The data on what would happen when nuclear warheads are used is rather sparse. The size of weapons used and the tragic lessons learnt at Hiroshima and Nagasaki do not extrapolate very well to a full-scale exchange of nuclear weapons in today's environment. The nuclear weapon technology and the delivery systems used during the 2nd World War could be termed as primitive today. What follows hereafter is an assessment of a situation where usage of nuclear weapons is not ruled out. It will be logical to base our calculations on factors that can be either controlled or predicted. We could also make pessimistic assumptions where control or predictions are impossible. An effort has been made to broadly study 'matters material' when one has to take into account likely deployment of a nuclear weapon.

Enough has been said about a nuclear battlefield, both in the erstwhile USSR and USA. Both the super powers had carried out examination of all aspects of a situation (where nuclear weapons get used) during the height of the cold war. They both had made plans for ensuring survival of as many people and as much military equipment as possible when the chances of exchange of nuclear weapons became very real in early 1960's. At one time the Americans and their NATO allies were worried about a Soviet first use. A large number of studies were therefore conducted to find ways and means of saving as large a segment of the population as possible. The Office of Technology Assessment (OTA) in its report to the government "The Effects of Nuclear War"[1] looks at several different kinds of attacks on the USA. It appears that the theme was to examine " second strike capability", ability to absorb initial shock, regroup forces and equipment and defeat the enemy. Several different scenarios are explored in the above-mentioned report and in other literature including recently declassified "Pentagon Papers". Estimates are given of casualties, damages and chances of recovery. The OTA report is usually referred to while laying down various parameters as prescribed in " The Effects of Nuclear Weapons"- a US Department of Defence and Energy publication. The available literature confines itself to a total nuclear war that was likely in the 'cold war' era. The option was not, per se, ruled out. One theme runs through out official and academic studies. That is, limitation of a reaction time and likely effect of nuclear weapons on the morale, equipment in the hands of troops and their physical well-being create a scenario requiring a very detailed examination. In our context it will be more relevant to study the implications of usage of nuclear weapons in a limited way. It will also be appropriate to examine the subject where one is reacting to a "first usage" by a likely enemy.

Conventional weapons cause damage to equipment, buildings and injuries and death to human beings. Their effect is limited to an area and depends on the high-explosive content of the projectile. By contrast a nuclear weapon delivers its damaging effect in three different and distinct forms- a blast, thermal radiation and nuclear radiation (a phenomenon that is confined to nuclear weapons only)-each of these may result in devastating effect on both man and equipment over a wide area. Another facet of modern (nuclear) warfare that can be directly affected is the damage caused to virtually all types of electronic devices and equipment by the electromagnetic pulse generated when a nuclear weapon is exploded. There is a pressing need for better understanding of issues involved (where logistics support to the troops is concerned) if we are to make intelligent decisions and workable plans about how to deal with the threat of a nuclear weapon being deployed by a potential enemy. These issues broadly will be:

(a) **Can the defence be deployed in such a way so that it is able to meet both, nuclear and conventional attack?** The defence in a future war will be confronted with problems that were not there in the past. It could be dangerous to effect a large concentration of men and equipment in a small area thus making execution of a tactical and therefore material support plan that much more difficult. Earlier, the warfare in the depth was limited because the firepower and mobility had been restricted. Now, the attacker has the option of using nuclear weapons to ensure heavy damage to the support systems. Cutting the supply lines will bring the operation to a successful conclusion in favour of the attacker in due course of time with limited damage to the power making first use of a nuclear weapon. To compound the problem the defence can never know in advance the size or number of weapons targeted on a particular location. It cannot make elaborate plans to minimise damage caused by a nuclear weapon.

(b) **Do we have the structure of our ground forces organised for fighting conventional war and suitable for nuclear war?** Our tactical doctrine requires appropriate mobility, integrated firepower and an ability to communicate. Threat of a nuclear weapon will demand wide dispersal of troops and their speedy concentration at the critical point and time even when the warfare is still conventional. This means heavy requirement of vehicles for fuel, ammunition, and rations. This also means heavy requirement of spare parts for keeping aircrafts, stores and troop carrying vehicles and tanks moving. We must also remember that communication and artillery's target surveillance equipments are sensitive to the effects of nuclear weapons. They also require sophisticated repair, supply and maintenance organisation. What is worse, this type of equipment gives away its location through electromagnetic radiation.

(c) **What role can tanks and air-borne forces play in such a situation?** (According to the Russian manual Vremya i Tanki " tanks are more suited than other types of military equipment to combat actions where nuclear weapons are used. In particular, they are suited to enduring powerful dynamic loads. The advantage of the tank is that its armour protects the crew against light radiation and decreases the effect of penetrating radiation, while the tank's actual weight gives it stability that protects it against the shock wave. As a result of this, the use of tanks under conditions where nuclear weapons have been employed makes it possible to wage active combat action immediately after the nuclear explosion on contaminated terrain.[2] All over the modern world armies are now dependent on transportation and communications networks and their attendant infrastructures. Communications and combat surveillance will be affected by the sensitivity of these equipments to nuclear weapon effects. This may restrict employment of such forces unless suitable measures are adopted to safeguard troops and communications means even when plans are made for conventional war. This means that the troops operating in the combat zone should have protective clothing, tanks be given nuclear protection and plans should exist for dropping of air-borne forces at re-designated places assuming the enemy uses his nuclear devices.

(d) **Is it possible to switch on the battlefield from conventional to nuclear war**? With the immense extension of the battlefield in the depth it will be difficult to keep the troops supplied. Unfortunately use of nuclear weapons by the opposition will virtually prohibit the necessary logistics support, primarily due to the fall out of a nuclear weapon and the damage caused to the roads and bridges in the operational areas. Thus, in case of nuclear weapons usage mobility may be denied due to lack of fuel and spare parts.

(e) In so far as a logistician is concerned there will be one question of specific relevance. **What significance will warfare in the rear (where most of the base and other facilities will be located) have in the future**? This is important because the enemy's nuclear reply may be directed not against the battle zone but against the far rear. (In an article published in The Soviet military journal the "Military Herald" while discussing logistics in a nuclear environment, it is pointed out that the dependence on rear services will in future be even greater than in the past; victory is in fact utterly unthinkable without a highly organised rear. Both sides in a future war will therefore aim from the very beginning at paralysing the rear of the opponent and protecting their own by organising reliable anti-aircraft, anti-paratroop, anti-atomic, anti-chemical and anti-biological defence.)[2] Deployment of material support will, therefore, have to guard against sudden introduction of nuclear weapons.

(f) **When is the stage reached at which one side can be expected to use nuclear weapons and what can it hope to gain therefrom at that stage?**

(g) **What factors influence identification of a suitable military target for deployment of a nuclear device?** To effectively and speedily conclude an operation suitable military targets have to be identified. Any troops concentration would always form an attractive target if a power wishes to employ nuclear weapons in a situation where too much is at stake.

Our past experience shows that both India and Pakistan have never (even by default) targeted a civilian or industrial location. It will, therefore, be reasonable to confine this discussion to military targets. Military targets; besides troop concentrations, could also be defence oriented industrial establishments. The situation would obviously cover a fairly wide spectrum of our population, cities, industrial units and centres.

Effects of a Nuclear Blast

Before discussing specifics of the issues involved it will be appropriate to broadly describe various consequences of a nuclear weapon, both on men and equipment. The energy of a nuclear explosion is released in a number of different ways:

(a) **An explosive blast**: When a nuclear device is set off the immediate result is an enormous amount of heat produced in less than a millionth of a second. This heat not only produces its own direct thermal effects but it also produces the blast and gives rise to the radiation. The blast is qualitatively similar to the blast from ordinary chemical explosives, but it has somewhat different effects because it is much larger. The magnitude of the blast effect (generally measured in pound psi) diminishes with distance from the centre of explosion. It is related to the height of the burst above ground level, and its effects are dependent on numerous other considerations that are common with any heavy artillery shelling in an area of operations. However, nuclear weapons, particularly tactical nuclear weapons[3] (a plea for distinction between tactical and strategic weapons was made in "On Limiting Atomic War", published by the Royal Institute of International Affairs. Eventually and with a view to make operational plans a distinction between strategic and tactical nuclear weapons was made), have a much more devastating power as compared to conventional artillery.

(b) **Direct nuclear radiation or INR (Initial Nuclear Radiation)** A nuclear explosion produces energy in three forms: light, blast and initial radiation. This is to be distinguished from the 'fall out radiation' which results when the radio active particles (the INR) produced in the explosion are deposited on the material which the fireball contacts when it touches the earth. Safeguarding the troops and equipment is directly related to the "Protection Factor". This is a measure of how much the incoming radiation is reduced by matter interposed between the radiation source and the eventual receiver; a protection factor of 10 reduces the radiation by ten. (An assessment and formulation of battlefield radiation casualty criteria was made by the U.S. Army and approved by the Joint Chiefs of Staff in 1975. Refer "Effects of Nuclear War, Journal of the Federation of American Scientists, 34 No2 1981")

Radiation effects on the human body were described in terms of three major syndromes:

- **Central Nervous System (CNS) Syndrome**: The threshold dose is about 200 rads.[4] CNS is usually visible within minutes of exposure, and, depending upon the dose, the symptoms may not be immediately observed. The symptoms range from apathy and drowsiness to convulsions and collapse. These two are characteristic of doses above 5000rads and lead to death within two days.
- **Gastrointestinal (GI) Syndrome**: The threshold dose is approximately 500rads. An initial period of radiation sickness which may last as long as three days normally precedes the appearance of the GI syndrome symptoms which are usually first seen from three to five days following exposure. Typically symptoms progress from loss of appetite and vomiting to severe diarrhea, high fever and coma. Doses over 1000 rads lead to GI syndrome-related deaths and usually occur within two weeks.
- **Hematopoietic Syndrome**: The threshold dose is about 100 rads. In this case, the blood forming organs get damaged, principally the red bone marrow. The prodromal phase of radiation runs its course, normally ending on the third day. There is then a so-called latent period during which no symptoms of radiation sickness are evidenced. The hematopoietic syndrome symptoms are usually first seen two to three weeks after the exposure. Typical symptoms are chills and fever, headaches and easy fatigability which is the rapid exhaustion following exertions. Few deaths can be expected due to the hemapoietic syndrome for doses below 200 rads. With higher doses, the more severe symptoms are evidenced such as purpora, which is bleeding under the skin resulting in patches of purplish discoloration, and bleeding from the gums. Deaths from this syndrome occur from three weeks to two months following irradiation.

(c) **Direct thermal radiation**: most of which takes the form of visible light. It occurs at the time of explosion and can be very intense, but its range is limited and the effect largely is related to the size and nature of the bomb. Intense pulse of heat and light from a nuclear weapon causes 'flash blindness'. This is rarely of a permanent nature. Troops may be dazzled at ranges of tens of kilometers even from quite low yield weapons. They may therefore be affected by their own side's tactical nuclear strike. It is essential that they be given warning of this as far as possible so that necessary protective measures can be taken.

(d) **Pulse of EMP**- Electro Magnetic Pulse (EMP) is a short intense pulse of radio frequency energy, originally called "radio flash", which is energetic enough to cause upsets or damage to most electronic systems. EMP does not originate in the weapon itself. It is a consequence of the interactions of nuclear radiation with the atmosphere. EMP has no effect on human beings but has catastrophic effects on the electronic equipment being used by the military. The only materials that are as sensitive as man to nuclear radiation are the semi-conductor devices of modern electronics, and certain optical systems. Optical systems will be at a risk from thermal effects which may cause crazing of lenses, charring of non-reflective internal coatings and damage to filters and photo-cathodes. Long fibber-optic cables are immune to EMP effects unless they have a metal strength number. *Nuclear scientists think that EMP probably presents the greatest threat to both civil and military systems as virtually all of them depend on electronics. A single atmospheric burst can have a countrywide effect as the EMP from a single high-burst can disrupt an entire national power or communications system. Equipment can be hardened against EMP, but the level of protection tends to degrade with time. Therefore, high standards of* *personnel training and equipment maintenance are essential to maximise the chances of our systems surviving EMP. When a warning of a nuclear strike is given, or when the probability is high, certain simple measures will greatly reduce these dangers. Upon receiving the warning, antennae should be taken down or disconnected, cables unplugged from communication vehicles and equipment and doors, hatches, windows in vehicles and buildings checked for proper closure. These steps will go a long way towards preserving the equipment.* The simplest as well as the most obvious way of protecting a weapon or communications system against EMP is to put it inside an electronic shield called a 'global shield'. This may be made of any reasonably good conductor such as steel, aluminum or copper, and need not be very thick-3mm of steel is more than adequate. However, it is essential that it should be free of gaps and apertures through which EMP could enter.

(e) **The creation of a variety of radioactive particles**: which are thrown up in the air by the force of the blast, and are, called radioactive fallout when they return to Earth. Their effects will depend on the distance such particles travel in the atmosphere and for how long they remain in the air charging atmospheric conditions with their radioactivity. These particles could also start a contamination chain of their own, thus causing an even greater damage to both men and material.

The distribution of the bomb's energy among these effects depends on its size and on the details of its design. How would the weather behave at that time? Local wind conditions and especially the amount of moisture in the air may make an enormous difference in the number and spread of fires that will invariably result after a nuclear bomb explodes. Similarly, winter, summer or monsoon months, usage of the bomb in a particular kind of terrain will also be contributing factors. The capability of own forces to absorb initial shock and damage will largely depend on the "warning time" and measures adopted to reduce the damage. A very high level of training, availability of ruggedised equipment, suitable protective clothing and ready provision of appropriate medical cover will play a big role in minimising causalities.

Before we examine the logistics and how will a nuclear environment affect material support it will be useful to review the kinds of operational situations with which a material support system might be concerned. Bigger problem is that there are far too many uncertainties for a reasonable and workable plan to support a military operation in such a scenario. Any effort to look at the entire range of effects compounds them. Mainly the uncertainty has to do with the degrce of protection available to the troops. Would they get enough warning to make use of best available shelter? How would they react to the situation? How fast will they be able to regroup for countering the effect of nuclear weapon? These will be some of the immediate concerns of the commanders. In numerous exercises carried out by the NATO forces a lot of stress was put on the requirement of troops to operate continuously at a high level of efficiency without depending on regularly organised lines of communication. An acceptable solution was seen in organising the forces into smaller self-contained units. But the overall logistics requirements were found to be the same regardless of how large or small individual units were expected to be. It was also stipulated that the mobile units must be able to carry all their supplies and maintain their equipment, independently of communication lines. The problem is that with more small and compact units there will be a greater need for re-supplies and maintenance. Fuel would be a much bigger problem because a lot of fuel destined for delivery may be consumed by the vehicles of supply services if they have to trace and chase the mobile forces for any indeterminate length of time and space. The use of aircrafts and cross-country vehicles for moving supplies might make the lines of communication less vulnerable but the vulnerability of base installations will remain. To withstand a nuclear strike, absorb its military consequences, restore equipment to serviceability, and restore the health of troops will be some of the key issues a logistician will face. Following situations will arise:

Initial Attack- Defence against a smaller nuclear power is much more feasible, although it is important not to overstate the likelihood of achieving a highly reliable leak-proof shield against such a strike. Sensors with capability to discriminate between real warheads and simple decoys will be required, at least within the theater of operation. Many different kinds of attack can be envisaged. These range from an attack of just a few weapons spread over a period of weeks to a massive spasm in which most of the nuclear weapons in the arsenal of the adversaries are detonated in a short period of time. A nuclear weapon, in addition to blast, heat and radiation, would cause wide spread fires due to ignition of grass, scrub and wooded land. Fuel and ammunition dumps, camouflage nets vehicle canopies etc could be easily set alight as these would have been heated and dried by direct thermal radiation. Immediate replacement of damaged material will have to be undertaken.

The commander on a nuclear battlefield will have yet another situation (where troops are concerned) to face in the form of ionising radiation that is inevitably associated with nuclear detonations. The wounded produced by the heat and blasts of the explosion are plainly visible immediately. Those who are affected by radiation may not be immediately manifest. Clearly, if the individual soldier is the determining quantity on the battlefield, then in a nuclear environment command concern for his survival must be greater than it has ever been. In a conventional war medical services are considered to be adequate because, in theory, evacuation is only an exercise in logistics. Based on past experience etc arrangements are made for the anticipated casualties, their transportation and other facilities organised. With nuclear weapons giving another dimension it will be a serious problem for commanders to find sufficient number of the medicos. There will also be greater emotional stress in such a situation where the individual may lack close supervision and group identification where even remote impact of a nuclear weapon could lead to disorientation of an individual soldier. In a nuclear war the Army may not have the luxury of time to make personnel including doctors, available to handle all kinds of casualty problems realistically. For the moderately and severely wounded, early and effective handling on the field and disposition in the medical channels will give a reasonable chance for survival to most. Care of those who are not seriously affected will be given priority within the operational theater. This will help in their speedy return to duty and will avoid unnecessary depletion of forces. In some cases the casualty will achieve restoration to a degree compatible with a situation where they can be reassigned duty at a future date. Therefore, it might not be a bad idea to include "nuclear first aid" as a training subject at all levels. Besides making help available immediately to a casualty such knowledge will be good for the morale of the soldiers. It will be better if medical units in the forward areas have attached helicopters as well as tracked and wheeled ambulance vehicles. Returning logistics transport, where situation permits, can be used for evacuation of the casualties. Traffic control organisation must be capable of diverting returning vehicles to "pick up" points according to the operational situation.

Post Attack- Before, during or immediately after an attack, those soldiers/workers who can will seek protection in blast and/or fall out shelter. The time period for which sheltering is required may last from days to months. This is the time during which fallout decays to a level permitting persons to spend significant amounts of time outside of shelter. The immediate post-attack environment is the one with which most planning of material support will be concerned. Besides the troops civil population in the vicinity of a nuclear explosion will also be affected. That means we have to identify potential targets and take measures, which will help in minimising the effect. The effects could be too wide spread and damaging to the people who are in the vicinity of such a weapon. They may not have any direct or indirect involvement in the on going operation. Other countries that have apprehended usage of nuclear weapons by likely adversaries started taking adequate measures to safeguard their population from a likely "first usage" by their adversaries. China carried out its first test in1964 and immediately went into creating a defensive environment for its population. The Chinese proudly display their tunnels that they have built so as to protect people from the effects of a nuclear weapon. Doctor Ishwar Ojha, a political scientist of Boston University, USA who visited that country in 1968 states:

> "So they took us to the busiest marketing section of Peking (now Beijing) and into a large departmental store. And on one counter they pressed a button; the counter rolled away, and we saw steps leading down. We went down about eight meters and found tunnels, and all well built brick and concrete tunnels, miles and miles of them. The entire city of Peking, they told us, has tunnels under it, with an entrance from every department store, every apartment building, and every residence. Inside the tunnels we saw kitchens, running water, sanitary facilities, food storage, medical facilities, all ready for use. In the event of a nuclear attack, they said, Peking's (now Beijing) seven million people can be safe in the tunnels in seven minutes, and can walk through them to twenty miles outside the city. And they told us that, since 1968 every major city in China had had similar tunnels built".[5]

Now that the threat does exist, may be it is time for us also to think on these lines and make a national level plan to counter a nuclear weapon. Keeping our environment in perspective where the prohibitive costs may deter us from undertaking an exercise of such a magnitude we should do what the Swiss have done by requiring all future construction to provide full shelter protection. Obvious necessities are underground food storage and medical facilities.

Post-Attack Recovery- Americans carried out nuclear tests in the fifties by creating special cities with human dummies. It was found that underground markets and basements in houses offered the best protection to those who are not caught near the ground zero. For those in the vicinity of the ground zero have no chance of survival.

As survivors emerge from shelters they will need to find uncontaminated food. Their ability to do so will depend on the magnitude of the attack, on their physical, physiological and psychological states. The degrees to which needed resources have survived (including weapons, ammunition, food and the means of reconstructing the technological infrastructure) and the possibly changed climate into which they will emerge. Each of these considerations is uncertain. The initial nuclear radiation is an immediate problem. The formation of fallout requires protection of the troops (and the factory workers) for periods of weeks to months. The fallout is usually placed into two categories: Early or local and the delayed (long range) fall out. Early fallout is that what gets deposited on the ground in the first 24 hours after detonation. This time limit is used since the fallout pattern is pretty well established after 24 hours from a single detonation. The following line of action can be adopted:

- **Find a safe and reliable source of water**. This is most essential because besides its requirement for drinking purposes water will be needed for cleaning of contamination from the body parts, decontamination of military equipment and putting off the fires. The human body can be decontaminated by washing with plenty of water. Filtration of water for drinking purposes considerably reduces the radio activity level.
- **Establish food supply system**. Immediately after a nuclear strike has taken place the survivors will need to be fed and reassured of availability of sufficient food. Tinned or packaged rations can be decontaminated by simply cleaning the outer surface. Plenty of these should be made available for distribution. Civilian

population in the area will also be affected. It will be normal to cater for their requirements as well so that regrouping of the fighting forces is not disturbed due to influx of refugees.

- **Establish health and care system**. Medical problems that can arise as a consequence of a nuclear strike have been enumerated above. Sufficient resources are to be made available to the medicos to deal with the situation. These resources need not be carried to the battle in the first instance, but should be available in the base and regional depots. Specific transport (including aircrafts and helicopters) should be earmarked for conveyance of these stores urgently when called for.
- **Establish search and rescue teams**. Devastation caused by a nuclear strike will cause a large body of troops to be staggered over a very wide area. These men will be totally disoriented and possibly suffering from some form of an after effect. Formation of suitable and well-equipped teams will considerably improve their chances of survival and restoration of their military skills that much faster. Trauma units could be created for dealing with more serious cases.
- **Establish task teams to deal with specific dangers like**:
  (a) Rodents and insects easily are the most dangerous carriers of diseases that lead to avoidable casualties. Disinfecting the affected areas will help in restoring the morale of local population and that of the troops.
  (b) Pilferage and theft of military stores. By enemy agents or by those who are hostile to the country. Organise local defence of the logistics units and areas earmarked for receipt of replacements arriving from the 'base depots'. A commander will be required to detail sentries and night watch of the concerned establishments. Meaning that much

more strain on his limited resources. We could muster our "ex-servicemen" for this kind of duty.

(c) Fire, rodents and insects. There will be a large number of devastating fires spread over a very wide area. Troops in the immediate vicinity of a nuclear strike, are not likelv to be in a position to do anything to put these fires out. Theater commander will have to earmark designated teams for this task. For speedy restoration of normalcy central pool of Manpower consisting of fire-crews, station health organisation teams Sanitation squads need to be created. Here again we can make use of our Ex-servicemen to beef up the teams so that speedy action can be taken to complete the task.

(d) **Decontamination**: nuclear fallout can be removed from most of the equipment by using a water-hose, scrubbing and using washing soap. For delicate equipment vacuum cleaning is the best approach. The entire range of weapons and equipment, which would have come into contact with nuclear particles, will need to be cleaned up in this manner because if left unattended these could create serious problems for the soldiers handling these and otherwise unaffected troops could get contaminated. It would not be a bad idea to create special decontamination teams who should also carry with them "dosimeters"[6] to measure radiation effect in a particular location and on a piece of equipment. Special protective clothing items should be made available to such teams before they start their task of decontamination. Sector reserves and stocks of relevant equipment can be held in the base depots.

Protection of troops: Thermal pulse from a low-yield weapon is so short that a man in the open has a very limited chance of survival. However, for a higher yield the pulse is long enough to allow a man to take evasive action like falling flat to the ground, rolling into a ditch or a trench. An all-enveloping cover (like a bunker or a shelter) will almost completely protect against thermal radiation besides reducing the effects of blast and nuclear radiation. Shelters can be of three types- (a) Maximum strength for key installations. Such installations will usually be 'operational headquarters', 'store (including fuel and ammunition) holding units', medical and allied facilities in the close proximity of operational areas. (b) Moderate strength for population masses in identified urban centers, factories of strategic importance. (c) Improvised shelters for small groups and family protection in residential areas in border districts. An effective shelter should be covered with at least three feet of earth. All shelters should have at least two readily available exits. Shelters will be also of two types/classes-blast and fallout. At a distance from ground zero where blast effects are not so serious the major problem is one of fallout. Different designs are called for to meet quite different purposes and the expenses in terms of manpower and resources will also be different.

Army Headquarters and depots directly under its control may be supporting troops who could land up in a nuclear situation. In such a case it may be more appropriate to allow the commander on the spot to requisition what he thinks is essential for his task in hand. Broadly, the material support pattern could be as shown in the following diagram. It will be seen that the overall structure gives a number of supply points with their inter-connectivity. Transport commensurate with the task in hand will be employed. It will depend on the distance between the supply points and the combat troops, type of terrain and the carrying capacity of the vehicles. Since consumption of fuel and ammunition will be unpredictable it will be prudent to earmark stocks(based on the past experience) to move to the locations where troops will eventually concentrate for their mission. The overall objective of the logistician will be to minimise the deficits in stocks of each commodity held with the combat units. To do so over an extended period of time will largely depend on availability of sufficient stocks at various echelons and our ability to replenish these stocks from our base depots and the time taken in doing so. We could create "nodes" to handle commodities and also set up repair and maintenance facilities in such a manner that it becomes possible to support various options available to the commander in a given situation. There will be limitations with regard to the quantity of a particular commodity that can be carried to the forward locations at a given point of time. In such a situation the higher commander can make allocations based on operational priorities so that all concerned are made aware of the resources available for a specific phase of an operation. When a nuclear strike is apprehended it will be desirable to keep high caliber ammunition and fuel sufficiently far behind in suitable bunkers or dug-outs so that "sympathetic detonation wave" is not created as can happen when all that atomic energy is released by the bomb . If such a situation is envisaged then it will be necessary that either adequate transport is earmarked for conveying this ammunition, rations and fuel etc and allotted to the formation as a firm commitment. Alternatively, "fit for air-dropping" ammunition is kept ready in a base depot to be dropped at a suitable place. All through the operations the main thrust will be to minimise the mismatch between supply and demand for combat units and monitoring of the discrepancy between planned and actual stock levels at all points in the system.

This discussion of logistics support is based on a deployment as shown in the diagram. Every variation and specific situation will lead to modifications and changes in the logistics support plan, but the core will remain the same.

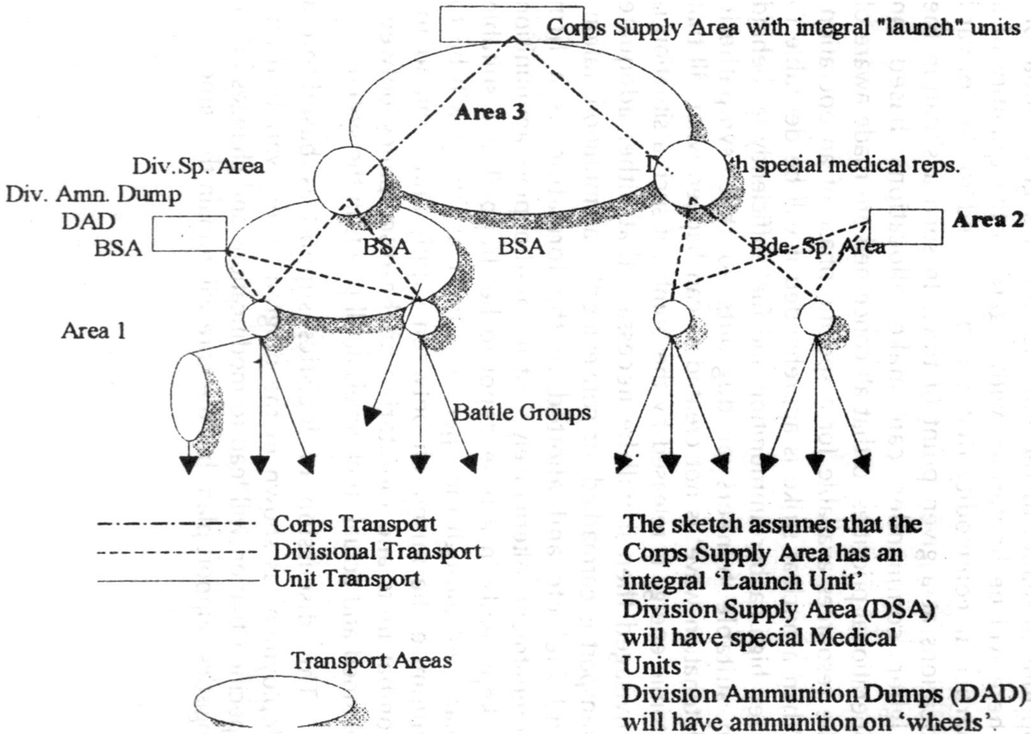
Corps Supply Area with integral "launch" units
Area 3
Div.Sp. Area
Div. Amn. Dump
DAD
BSA
BSA
BSA
th special medical reps.
Area 2
Bde. Sp. Area
Area 1
Battle Groups
Corps Transport
Divisional Transport
Unit Transport
Transport Areas
The sketch assumes that the
Corps Supply Area has an
integral 'Launch Unit'
Division Supply Area (DSA)
will have special Medical
Units
Division Ammunition Dumps (DAD)
will have ammunition on 'wheels'.

In a nuclear battlefield there will be a (pre-planned) quick dispersion and assembly of troops requiring speedy replenishment of expended ammunition and other stores. It could happen that there will be ammunition sub-locations of a "Field Ammunition Depot" already existing in the area of operations. In such a situation "Divisional Ammunition Area" need not be set up. However, it will be important to earmark adequate transport for carriage of ammunition to a unit location or a 'refilling point'. Correct mix of "armour ammunition", artillery ammunition" and other requirements like missiles and small arms ammunition should be kept ready for supporting those troops who would be either exploiting usage of a nuclear device or reacting to the enemy. As the medical and related problems are likely to manifest themselves in a big way it will be appropriate to allocate suitable resources at all levels for speedy redress of a medical requirement. Helicopters, tracked ambulances and those doctors who will help trauma cases should form part of the divisional medical set ups in those formations who are likely to attract the enemy's nuclear devices. Sufficient tinned ration, washing soap, water either carried or sources identified for washing of nuclear effect be made part of the logistics set up supporting the forward troops. "Engineer effort" for preparing bunkers and defensive positions be allotted from higher formation resources. They could also be tasked to prepare and man water points. and viable solution to our requirement in terms.

The troops fighting a conventional war when chances of usage of nuclear weapons exist must be so deployed as if they were fighting a nuclear war. We must train our troops to be self-sufficient and be able to operate with little or no direction. We must anticipate a situation where after a nuclear strike the communications could be by voice or couriers. That a likely use of nuclear weapon by the opposition will make the material support plan that much more difficult should be a major focus during planning and exercises. Once we automate our support systems Linear programming models can be adopted to arrive at a correct mix of resources including transport to arrive at the most economical of manpower and resources.

Protection against thermal radiation- Equipment

Unless a piece of military equipment is in direct line of a nuclear strike the equipment is generally protected against radiation and other effects. Military equipment is generally designed to survive high and low temperatures, lightning strikes, vibrations and shock, electromagnetic emissions from nearby radar sets and so on. Where the environment of blast, thermal radiation, and perhaps most importantly, EMP are concerned equipment vulnerability is generally an unknown phenomena. For example, tanks are assumed to give full protection to their crews against thermal and direct blast effects and a degree of protection against nuclear radiation. This particularly means that alerted, dispersed units and their essential support (e.g. warheads, intelligence, and logistics) should be survivable. However, the survivability of equipment may be affected by inadequate maintenance, or by modifications made without proper considerations of the implications for nuclear survivability. This will be particularly important where EMP hardness is concerned. It is therefore essential that those responsible for modifications should have not only a good understanding of nuclear -weapon effect, but also an awareness of the design philosophy employed by those who originally developed the equipment. We must understand that the conventional weapons continue to occupy an important place in the overall material-technological base of the Armed Forces. But they must be improved applicable to new conditions and new demands of warfare.

For a military success with minimum losses, when the environment could turn nuclear, there is a need for continuous and inquisitive search for armaments, better organisation of tactical grouping and reliable methods of administrative support. With nuclear weapons having become a reality a commander must have resources required for his task. To meet his logistics requirements special effort and expertise is needed. This is a specialist job. This requirement of military service imposes a scope that is unrivaled by any other enterprise. In fact, in today's environment and with the technological advancements being made in the field of armament and electronics there is a greater need for expertise and a higher level of professionalism in the services particularly when things like EMP could render the entire system useless. The modern war (where a nuclear device could be used) will be fluid, flexible, spread out and will consist of many battles fought by small units. Because, now our weapons, vehicles and equipment are much more complex and intricate they require more maintenance, a wide variety of fuel and spare parts. Meaning that there will be need for more intimate material support than before, during and after the operations. The tactical logistic problems will be very great requiring the maintenance of vehicles and weapons and the rehabilitation of these weapons on an even greater scale than in any previous war. The one able to solve these problems in peacetime will have greater expectations of success in war, be it conventional or nuclear. For ensuring success in any future war we have to reexamine our support systems. Our logistics systems are relatively unchanged. We rely on fixed depots, linked by road and rail with the forward areas. There is limited cross-country facility available to support services. In case nuclear option gets to be used the depots, rail and road communications will be easy targets. Congestion created by the survivors running from the affected areas will add to the problems. Mobile operations, dispersed formations and an extensive battlefield will call for faster and more accurate response from the support services. These demands will be further accentuated because supplies have to be kept in and delivered from dispersed depots. Thus the degree of mobility and dispersal of combat troops in the field will largely depend on the capability of the services to respond to a wide array of situations which have not been faced in an actual operation. This will also depend on the flexibility of the support services and their capability to switch successfully from the conventional to nuclear war. As it is the list of desirable improvements in our logistics services is long enough. There is a need for conscious management strategy that takes into account factors like: -

- Future army concepts, dealing with combination of defensive and offensive operations, and designing material support systems accordingly for achieving greater mass and effect. Unification of the support system, standardisation of our weapons and allocation of tracked vehicles and air effort for supply of essential stores and evacuation of casualties during exercises will help in making efficient support plans.
- Increasing reliance on information systems for coordination,
- Emerging technical opportunities, and increased integration of technologies with the civil sector so as to improve availability of spares, communication equipment of common usage etc.
- Ability to maintain capability in Army niche areas, such as armour design and penetration machinery, evolving better methods to absorb technology quickly and enhancing the survivability of tanks and their crews.

For a long-range appraisal of logistics functions and requirements our lens must widen beyond the immediate future because to build capability to react to a nuclear strike needs a lot of resources in terms of time, effort and money. It also entails preparing the civil population- because a nuclear weapon, contrary to a conventional projectile, has very wide ramifications. A theatre nuclear force R&D programme has to be initiated with the following objective:

To assess the survivability of various military elements under conventional and nuclear attack, identify deficiencies and develop improvements to develop technology to counter possible future threats to the survivability of these military elements in different theaters of likely military operations. We must be prepared to fight the war of nuclear parity. We need to develop organisational and tactical concepts that can be used by a unit or formation that can get along with the fuel, food, medicines, spare parts and ammunition that it can carry or forage enroute. For this we may need to appoint a "Logistics Readiness Officer" who will be charged with advising and coordinating the logistics readiness of the operational command. We must develop more rugged items that are also economical to operate.

Notes and References

1. *The Effect of Nuclear War*, published by the Office of Technology Assessment, USA.
2. *Conventional Warfare in the Nuclear Age* by Otto Heilburunn.
3. The British Minister of Defence defined in the House of Commons on April 3, 1963 a nuclear weapon as tactical if it is directed at a tactical target. In this view, it appears that a nuclear device be regarded as tactical if it is directed towards troops, their supply and transport facilities, airbases, seaports and infrastructure.
4. The rad is a unit of absorbed radiation roughly equipment to a roentgen, which is commonly used in measurements describing human biological responses to radiation exposure.
5. I Ojha, Survive Jan-Feb. 75, Vol. 8, Number 1, page 1, as quoted in *Survival and Peace in The Nuclear Age*, by Lawrence W. Beilenson.
6. Dose Rate Meter or Dosimeter is a device that indicates the total radiation dose. Safe limits have been prescribed. This helps a combat soldier to avoid radiation effect while he is engaged in cleaning an area where a nuclear device has been used.

11

The Siachen Challenge

The desert is said to be a tactician's dream and the logistician's nightmare. Siachen glacier is a nightmare for both. Siachen in itself is an area of extremes. It discounts normal human habitation by its unpredictable surface conditions, inhospitable climate and attenuated atmosphere. This is an area no man has ever inhabited- and only a few had traversed- before Pakistan and Indian troops moved in to wage a bitter conflict over fifteen years ago. Both the countries are spending up to Rs. 5 crores per day to assert their right to this 76.8 kilometres long glacier. On this Glacier, where nearby peaks reach 7,010 meters (23,000 feet) and temperatures drop to minus 45 degrees centigrade the frostbitten armies of India and Pakistan are engaged in a conflict that is bloody, and surreal. Here cold and crevasses kill more troops than the opposing armies do. Nearly every system in the human body gets affected due to stay at these altitudes. One of the biggest killers is pulmonary edema, brought on by the high altitude. The lungs suddenly fill with blood and fluid, leading to a slow gasping suffocation. A soldier can survive if he can be evacuated in a helicopter to a hospital in Leh, the capital of Ladakh, at an altitude of 11,555 feet. There the casualty is put in a huge iron pressure chamber that simulates sea-level pressure, the only way to reverse edema. When inclement weather grounds evacuation helicopters, as it often happens, the afflicted soldier usually dies. Other afflictions the altitude and atmosphere impose are frostbite and chilblain with potential to maim permanently or kill instantly if unattended to, in a span of few minutes. In this high, frigid air, skin bonds with metal, sweat turns to ice, and there is not enough oxygen to light even a matchstick. A man cannot carry more than 7 to 10 kilograms and cannot walk at a pace faster than 200 to 300 metres per hour. If he strays from one of the designated footpaths, a soldier risks being swallowed by a hidden, snow-covered crevasse. There is no room for brilliant tactical maneuvers and no local resources to exploit. The nearest habitation is 10 days away by foot. There are no trees, no grass and no wildlife, only an unending desert of snowy white. The 'Saltro' ridge is home to troops of India and Pakistan in its innumerable natural and man made caves and a wide variety of other habitation infrastructure which could be a combination of empty steel boxes, jerricans, parachutes to fibre reinforced plastic/glass huts.

In the absence of habitation of any kind and with only ice and more ice our commitments in " Siachen glacier" have thrown a special challenge to the support services. To keep combat troops supplied with their day to day requirements when they are engaged in fighting an enemy and the elements demands very high level of commitment from the logisticians. Since, no army has ever fought at those heights and in such an inhospitable terrain our support systems have evolved over the years based on experiences gained and dealing with numerous situations as they arose. It is not that the troops have not been stationed in high altitude areas earlier. They have been there in the NorthEast and places like Durbuk in the Ladakh region even earlier. Maintenance of these troops barely reaches satisfaction level even when the Indian soldier is known to be least demanding. It has not been possible to develop an 'all weather road' to Leh. 'Advance Winter Stocking' and air effort are the only avenues available to provide material support to the troops stationed in the region when heavy snows close down the roads. With the rail head at Jammu the haul by road to Leh is too time consuming and expensive in terms of money and other resources. Emergence of Siachen as an operational commitment added another big dimension to the (already strained) material support system.

All our earlier experience, logistics support policies and planning could not measure up to the demands of the glacier. The situation here called for far too refined and improved equipment, higher levels of technological competence and resources that could give our soldiers wherewithal to fight the enemy as well as the elements. The height of over 18000 thousand feet and temperature ranging from minus 70degrees centigrade to zero degrees is not an environment where any classical model of material support can fit. The 'deep freeze effect' solidifies everything in minutes except kerosene and rum. Guns freeze due to the climatic conditions. If you pick up a rifle without a pair of gloves the skin can peel from the fingers. This effect also kills appetite and the rarified atmosphere makes every kind of human activity very difficult. The troops in their puffy high-tech snowsuits tethered together, walk only on footpaths tested for ravines and crevasses. (Earlier at the beginning of the operation the troops were not adequately protected against severe winter conditions in the glacier, leading to an alarming rise in the number of weather related casualties. With acquisition of insulated pre-fabricated huts, special Swiss winter clothing and heating system the position has somewhat improved).

The troops occupy pickets and posts living in arctic tents and bunkers at heights ranging from 18.000 to 21,000 feet. These pickets are connected with telephones and in some cases radio communication is also provided. However, due to shifting of snow and melting of the glacier the cables snap easily. To repair and make the system operative again creates its own set of problems when the precise spot of breaking of the cable has to be located and the defect rectified. Where radio communications are concerned these are prone to disturbances and jamming by the Pakistan soldiers. Only "Cheetah" helicopters can reach these pickets with 20 to 30 kilograms of load at any given point of time. The per hour cost of a cheetah helicopter works out to Rs.80, 000 per hour. It takes several helicopter sorties to equip most Siachen posts with bare minimum essentials. According to one estimate transportation costs for the supplies are estimated at over Rs. two crores per day.

On an average, a detachment of four Cheetah helicopters can airlift up to 65 to 70 tonnes of load in a month. Similarly, a detachment of four Mi. 17 helicopters is capable of air dropping 300 to 350 tonnes of load in a month. By no stretch of imagination can this be called an ideal solution to the requirement if one considers the fact that every item required by the soldiers has to be flown in. This include precooked food including chappaties, medicines, ammunition of all kinds both for the personal as well as area weapons, spares for vehicles at the 'base camp' and for communication equipment through out the operational area have to be airlifted to the pickets. The magnitude of the task becomes clear when one appreciates the fact that the helicopters are flying much beyond their prescribed heights and tolerance limits to keep troops equipped. What brings in another element of uncertainty is the fact that air dropping of stores needs extreme precision and care because strong winds can easily take the stores either into the enemy territory or land them in a crevasse or some unreachable spot. For that matter the weather may not even allow these sorties to take place for days on end. There is no room for setting up conventional dropping zones either. The elements allow only barely adequate chance to the helicopters to provide sustenance to those committed to guard our frontiers at those heights and in such an inhospitable terrain.

Base Camp

The base camp where from the troops are supported is located at the foot of the glacier at an altitude of 11,400 feet and is 75 miles from Pratapur-the nearest formation. Army Engineers have linked this camp by an all weather road with Pratapur and Leh. One part of this road crosses the 18,382 feet high Khardungla pass making this road highest in the world. This base camp is a busy 'integrated' depot from which arms, ammunition and other supplies are sent to some 75 forward posts in dozens of daily helicopter sorties, mostly on bulky Mi-17 helicopters. For higher altitude missions, smaller Cheetah choppers are used. Troops are trained in this area; they acclimatize for six weeks while being taught skills like 'ice-climbing' and winter survival. Equipment includes Austrian Mountain boots and Swiss clothes. The base camp gets to have a crazy, skewed look because of 'pillaring'- an effect that leaves any covered area settled earlier' of snow several feet above ground during the summer. Heli pads, gun positions, even tents and fibre glass huts end up on snow pedestals and eventually have to be moved elsewhere with considerable effort. A bigger problem is garbage. In summer, last winter's neatly dug pits resurface in the summer melt down, inviting scavenging crows that attract the attention of enemy gunners besides creating usual hygiene problems associated with dirty surroundings. This base camp forms the only life line for the troops posted on pickets on the glacier. The effort necessary to supply more than 75 posts and well over a brigade of troops on the glacier and in the Saltoro is nothing short of monumental. Mainly, the problems relate to keeping troops supplied with warmth, ammunition of all kinds, rations, and above all maintaining their morale. The main supply effort sustains the fifteen years old camps that provide access across the Saltoro ridge and onto the Siachen. Because of the elements and continuous movement of men and material there is a constant need for a regular work force. Widely scattered villages are available in the lower regions. The army hires some labour force, both as porters and road workers. But bigger brunt of the load is handled by the troops themselves.

In order to appropriately understand the issues involved we will deal with subject commodity wise. First and foremost will be rations for the troops. Other than in the base camp and that too when the weather is not extremely severe it is not possible to provide traditional "freshly cooked" meals to the troops. They, perforce, have to make do with tinned rations. Tinned rations of appropriate calorific value are, either air dropped, taken by helicopters or man-packed. In addition, dry fruits etc are supplied for providing warmth. Generally these dry fruits are not a welcome addition because too much of it can upset stomachs and the temperatures are not conducive to provide relief to any soldier who has a bad stomach. A single precooked chapati costs approximately Rs. 50 when it reaches the troops at their posts. Even then the whole exercise of bringing this precooked food to forward posts sometimes reduces palatability. In the extreme cold conditions there is deterioration in the food quality. It could happen that bad weather does not allow sorties to take place for days. Introduction of refrigerated vans from Leh to the base camp can improve this situation to a large extent as freshness of the food can then be ensured for a much longer duration.

The equipment has to be given special treatment on Siachen. An Artillery gun is flown up piece by piece because the helicopters cannot carry a whole gun. The guns are thereafter assembled on the spot. Artillery shells do not subscribe to any theory of science and are seemingly freed from the normal laws of ballistics. Consequently, there is very limited accuracy and resultant damage to the enemy. Next thing to be considered is ammunition. The operations consume tremendous amount of ammunition, particularly artillery ammunition. Loads are heavy and the staff engaged in ensuring adequacy is always under pressure. Ammunition is being taken up to the "road-head" by road, is air-dropped where possible and conveyed by air-craft or helicopters to a certain extent. Thereafter it is man-packed because items like snow-scooters got caught in the normal "bureaucratic inertia" as the concerned sanctioning authorities probably had not as yet been able to make sufficient number of trips abroad on the premise of looking for suitable equipment. Now that some of these have arrived the organisation can take on the onerous responsibility to a limited extent. Main problem is with certain types of missiles, as they tend to be erratic due to solar rays and their impact on the missile electronics. Because of the rarified atmosphere and extreme heights carrying of small arms ammunition in adequate quantity is always difficult. Missiles tend to crack, ammunition if not consumed within a specified time frame may behave in an erratic manner, oils, lubricants and medicines may freeze. In such a situation our classical methods of material support are not likely to be very effective.

A major requirement is that of providing fuel (particularly kerosene oil) for warmth, for cooking and certain other activities associated with high altitude areas. Availability of a suitable landing strip as close to the location and laying of a "pipeline" thereafter taking the fuel as close to the troops will be a big help. (It is understood from press reports that such a pipeline has since been laid up to the base camp. Feasibility of using 'lift irrigation technique for carrying the pipe higher need urgent examination so that man-carry or air dropping of this load can be reduced. This will make more helicopters flying hours available for other tasks). Another big cause for concern is that of psychological and traumatic effects of isolation, and "ultra-violet" rays on human beings. Army doctors admit that soldiers returning from Siachen show signs of excess stress and symptoms of psychological disorder. It should be possible for the psychologists and psychiatrists to visit forward posts and provide mental relief to the troops. Cases of frostbite, other diseases associated with extreme temperatures and heights are not uncommon. Unfortunately, the organisational efforts to improve the situation often get thwarted because of certain unscrupulous elements who in their hurry to make quick bucks provide substandard equipment and accessories to the troops already under pressure from the elements. Recent newspaper reports mention about soldiers losing their fingers due to frostbite because the gloves supplied to them were not adequately treated to give "thermal protection". Our "Quality Assurance" agencies must gear themselves up to deal with the contingencies of more sophisticated and modern equipment being inducted into service at the glacier. Their inspections must be total and thorough. Such lapses that led to an avoidable loss of limbs due to substandard gloves need to be ruthlessly punished so as to prevent adverse effect on the troop's health and morale. Visits by psychiatrists and counselors will also be a big help. The troops do not have adequate and appropriate surveillance equipment. This drawback negates all the sufferings being undergone by our soldiers. They should be immediately supplied with telescopes, spotter scopes and thermal imagers in increased number so that can be more effective in observing enemy movements.

The glacier is a nightmarish challenge. One cannot change the environment. But innovative methods for keeping troops warm, well fed, supplied with their needs can definitely improve the situation. "Rope ways" with intermediate stations can provide some relief to the troops as the ropeways can be used for moving stores including ammunition and medical supplies. Where possible troops should also use this facility in order to conserve their energy. Because of the rarified atmosphere the helicopter can not take more than 25 kilograms per sortie. Efforts to make 'heli-pads' of bigger specifications can be made. Helicopter load can be commensurately increased with the consumption of fuel per sortie to the forward posts. In the meanwhile, Arial rope ways and snow scooters are our best bet. As per a report published in the 'Jane's Defence Weekly' there is a proposal to link high altitude posts on the glacier via a chain of 'ski lifts'. Making of such an arrangement will make taking of loads of all kinds forward more reliable and effective. Induction of "Yak" pulled carts can be another way of ensuring better maintenance. Operation of reindeer pulled sledges on the "Saltro Ridge" is another experiment that can be tried. Repair and maintenance of equipment, particularly communication systems is an area of urgent concern. Efforts should always be made to repair such equipments "in situ" because pulling them out for base repair etc. will always pose special difficulties.

Similarly, feasibility of setting up of "bio-gas" plants can be examined where human waste instead of creating unhealthy environment is put to use for creating heat and energy. Windmills where the wind velocity is not unusually high can also be tried for generating electricity. When the weather is clear there is enough of solar energy. This can be harnessed and used in solar cookers, heaters and lamps.

For improving the communications between the base camp and forward posts optical fibre cable should be installed at the earliest as a durable communication link is vitally important in this area. VSAT link can also be provided. This will be faster but prone to enemy interference and may have some security problems. These problems are normal in any communication network and can be resolved by evolving suitable drills and practices. The morale and confidence level of the troops serving on the glacier will go considerably higher when they get reliable means of communication. For then, besides being in touch with the base, they will also be able to report enemy movements much faster for reaction at an appropriate level.

Siachen is the longest air mobile battle that the Indian forces have been engaged in. From the initial deployment in April 1984 to subsequent redeployments and thereafter, continued maintenance of the troops at the highest battlefield has been the task of the Indian Airforce's helicopter fleet. Their training, dedication and devotion to the task is unmatched. But they cannot always overcome the elements. It is for the logisticians to come up with solutions to the problems that the glacier throws up with menacing regularity. Logisticians must become aware of the weaknesses and strengths of the system and develop means to overcome these weaknesses. It should be possible for the support services to give commanders more inputs to evaluate more and better options and an opportunity to react to a situation in an appropriate time frame. One can use analytical techniques and inventory models for working out an effective support system. But the situation defies any such analysis. The support system will perforce be complex, composed of several locations and agencies that will be responsible for a wide range of activities. Because of the nature of terrain, uncertain weather conditions, with extreme cold and heights testing man's survival skills it will be important to specify inventory levels and repair back up at several locations that will maximise total system effectiveness. For ensuring speedy return of equipment to serviceable condition we will have to consider (a) variable repair capability of each location, (b) personnel skill levels and capabilities, their availability and utilisation as close as possible to the location of equipment casuality. Medical authorities could set up 'Trauma Centers' at the base camp so that troops who suffer from any climate related problem could be given immediate help. This centre will be particularly useful when ropeways and ski lifts become operational and it becomes possible to evacuate a casualty to the 'base camp' rather than wait for a helicopter that might not come due to bad weather. Simulations and related techniques in the field of Operational Research can be applied to the management and organisation of logistical systems of any complexity and size. These techniques provide a flexible and analytical tool that can be used to determine the most effective support for a given logistics system. Main advantage will be that such an evaluation will provide an ability to continuously evaluate alternative logistics policies in light of experience gained over a period of time. In addition to enhancing effectiveness such an analysis will also bring in sizable reductions in the costs of maintaining troops in such surroundings while enhancing the 'level of assurance'[1]. Whatever simulation model we select it should have the ability to simulate increase and decrease in maintenance personnel giving a description of the maintenance capability at each intermediate location in terms of the number of maintenance personnel and their respective skill levels. The lowest skill level for repair of each high-value part be also identified so that commanders know the stage to which an 'out of action' equipment is required to be back-loaded. This means that a detailed description of the logistics network will be a desirable factor for enhancing our efficiency where support to the troops in Siachen is concerned. These inputs, in addition to specific peculiarities of a particular region when put through the simulation models can give us maintenance costs versus inventory investment for various options along with the risk level in terms of loss of functional abilities and fighting capabilities of the soldiers at the glacier.

Note

1. RAND Corporation in USA have developed two such models - LASS (Logistics Analysis Simulation Systems) and METRICA (Multi-echelon Technique for Recoverable item Control).

12

Our Wasteful Ways

> It appalls me to think how many failures occur in this very last link of the logistics chain. Equipment is manufactured at great expense. It is shipped 5000 miles by train, ship and truck. It is issued to the troops and eventually, with great labour, carried to the top of a mountain. How many times, at that last point, has this whole enormous effort been thrown away, as carelessly as a burnt match, by the happy-go-lucky negligence of the very people whose lives depend on keeping the stuff in shape?
>
> "Commanders Must Know Logistics," Lieut. Gen. W B Palmer, U S Army

'History', wrote Richelien in his political testament, 'knows many more armies ruined by want and disorder than by the efforts of the enemy.' Logistic discipline is attained when we train our troops to realise that the community has undergone some sacrifice in making those things available to the 'frontline' soldier without which he may not be able to either contain the enemy or the elements. One has always been surprised to see commanders wanting the "authorised" coal to be burnt whether the weather permitted it or not. Or sanctioning consumption of a particular item regardless of an immediate need for the same because if the expenditure is not incurred during the financial year the scales could be reduced. The item could be even taken off the list of entitlements.

Some how, and wrongly, an impression has come about that materials required by the Military regurgitate from a bottomless pit and there is no real need for the people to be careful and considerate in expending our resources. Often enough, and in the guise of operational requirements or in a bid to keep the morale of the troops high, considerable wasteful expenditure is incurred by the military. A lack of discipline is also evidenced when a consumer attempts to compensate for lack of foresight (by not placing a demand for his requirements well in time) by means of specifying an unwarranted priority, an unrealistic delivery date, or by sending reminders and expediters. It has been seen that certain units have got the "demand forms" printed listing the entire list of stores that are required to be demanded at fixed intervals. The form shows " quantity authorised" and the "quantity demanded". Invariably, the quantity held is shown as 'nil' so as create some kind of pressure on the issuing agency. To compound the problem the demands are generally marked "Op-Immediate" even if the demand form contains a vast range of "cosmetic items". To further add to this perceived urgency far too many hasteners, including couriers, are dispatched to the depots at considerable expense to the state for obtaining such stores. In some cases the quantities held on charge are expended quickly so as to have a 'nil' balance on hand. Worse is the case of a large number of combat and combat support units. In their 'Equipment Tables' the "quantity authorised" is shown "as required". This is taken as an official mandate to be indiscreet. Certain units consider it to be a blank cheque for demanding hundreds of litres of "Paints"- "Oils and greases" and many other such items that are intended to be used only for maintenance and upkeep of the equipment. Often enough specific quantity required for the purpose (per piece of equipment) is prescribed. However, since the operative clause is "as required" there almost always is a mad rush to collect as much stock of these items as "a busy in building unit stocks" quartermaster can. This is a mindset that needs to be shed off as quickly as we can, because these wasteful activities are not doing any good to the Military. Such practices put the whole procurement cycle out of gear besides causing functional and administrative difficulties in the service units.

Such hoarding of stores cause problems in the holding units as well; particularly when they have to move on short notice and enough transport is not available for carrying every thing. It is important to educate all concerned that all stores whether expendable or otherwise are community property and are required to be used with utmost care and economy.

Technical Inspections One 'major' stumbling block in efficient use of resources is the requirement of "periodic technical inspection" of equipment and vehicles in the hands of troops. These inspections are essential for ensuring availability and serviceability of equipment. Problem is that with our basic approach to have more of "spit and polish" with limited concern for fitness the technical inspectors want the vehicles to be freshly painted, gleaming and well polished when presented for inspection. This means painting of every vehicle every year before its annual inspection. If we take the number of "load carrying" and other vehicles held by the Army a huge quantity of paints is required to be made available to all the units for this purpose. (A three tonne vehicle requires over six litres of paint for one coat of painting; Artillery or an Engineers' regiment is authorised over 100x 3 tonne vehicles alone. If we take all the vehicles, including and besides load carriers that are required to be painted every year for their technical inspection the magnitude of this wasteful exercise makes one feel uncomfortable.) Since this huge requirement cannot be procured centrally because of the amount of money involved and inability of standard and reputed firms to meet this heavy demand local purchase of paints, particularly 'olive green ' paint and thinner etc. is resorted to. The quality of such purchases is often doubtful and wasteful because officers who hardly have expertise to assess the technical specifications of the product they are making the state to pay for usually buy these paints. Problem of these substandard stores finding their way into Army is more real in remote areas where regular and reputed suppliers are not available. More often than not usage of substandard items do more harm than good to the equipment. The concerned units want only to see them through what is known as the "CEME Inspection". They are not particularly concerned with the quality of the items supplied. And even if they are, there hardly is any other option because the inspecting agencies might take "non painting " of vehicles as a slight and thus give an "unsatisfactory" report making life difficult for the commanding officer and his "Military Transport Officer".

Secondly, these technical inspections and "painting " requirement etc. need vehicles to be " jacked up" for a considerable length of time creating all kinds of problems for the concerned unit. Collection of rations, training of troops, essential movements, everything suffers in the process.

When no military officer, certainly from the support services, gets his private car painted every year reasons for expecting the government vehicles to be thus treated are not clear. It is not as if this annual painting of the vehicles enhances either the life or performance of the vehicles concerned. Millions of rupees in the form of "paint procurement", thousands of "man hours" in carrying out this activity and still an uncertain serviceability is an avoidable drain on our resources. It is time formation commanders insisted that functional efficiency of the equipment in the hands of troops is more important for them to assess 'operational readiness' of a unit than the cosmetics.

Disposal of Salvage Stores

We inherited from the British the system where a "condemned item is returned to the nearest "salvage store section" of the nearest Ordnance depot. This means that a unit located in "Dehradun" or any other location will pack an unserviceable battery (or any other military store that has run through its serviceable life), take it to the nearest railway station, and send it by rail (or may be by road) to the Ordnance Depot at "Allahabad", or any designated Ordnance depot. The depot, in turn, will collect a large number of such batteries, and all the tattered tents, blankets, MT spares and such like items, hold them for a considerable length of time and account for them by employing huge manpower, land and other resources. Eventually these items will be auctioned as scrap. The "revenue" thus generated is taken as an "asset". Costs involved in sending the unserviceables to the depot, the salary and other expenditure incurred in employing the work force for the task, and various other costs usually associated with "serviceable inventory" (but spent in looking after the scrap material) is not taken into account. (As per a study carried out by the Army Headquarters in any one year Rs. 15 to 16 crore were spent in backward movement of discarded class 'v' vehicles alone. Approximately another Rs. 10 to 12 crore were spent for back-loading other unserviceable stores to the salvage depots.) This amount does not include the cost incurred in dispatching unserviceable stores and vehicles by road, either to an Ordnance depot or to a 'rail-head. If we take the total perspective of this wasteful exercise the magnitude manifests itself demanding an immediate corrective action. A unit located in or close to a town could very well return the 'unserviceable' battery to the "authorised dealer", get discount for the same and obtain a new one across the counter. There are many such items in our inventory where "buy back" facility is available. The change in our procedures where such transactions are made operative could mean genuine saving in the form of rebates from the dealers. It would also give faster replacements to the users. Similarly, unserviceable blankets etc., which often are given to registered charitable institutes after rotting in the depots for months, if not years, could be disposed off locally by commanders as there are any number of "poor people homes" and such like located in their close vicinity. In so far as unserviceable vehicles are concerned these can be disposed off by carrying out auctions or tender under the authority of a station commander. Only those unserviceables be sent to salvage which could either compromise security or cause problems if they fell into wrong hands. These will normally be armament spares, signal and wireless equipment and certain other high value items. We could save on manpower employed for manning the 'salvage stores group' by cutting down their workload. This line of action would allow the poor to make use of the items when they still have some kind of warmth and utility. Clearly, the whole of our salvage disposal system is such a wasteful exercise that a pragmatic decision and change of procedures is urgently called for. This could bring in a saving of, at least, one percent in over all military budget.

To be fair, items like "broom sweepers" which were earlier Ordnance responsibility and, therefore, needed packing and dispatching like any other item have since been taken off the inventory. But it took years to convince the Ministry that the packing and dispatching cost in these cases was much more than the cost of such items.

Technological Adventures

Because we have always treated defence and development as rivals for the country's resources we have pursued technological growth in both the fields in a compartmentalised fashion. In our hurry to catch up with more advanced West a large number of high technology projects are undertaken. Perhaps, the most wasteful is our tendency to try indigenisation of a very large range of items, sub-systems and systems whether technological capability exists or not. With a view to project ourselves at par with more advanced countries of the world we have spent a lot of time, energy and effort in trying to develop highly sophisticated "advanced jet trainer" and such like equipment. We did not create sufficient infrastructure, neither in the civil nor in the defence industrial sector before undertaking such 'high technology' ventures. We have had ambitions to develop sophisticated supersonic aircraft without having a credible civil aircraft industry. We have been trying to develop a variable compression engine for the armoured tank without bothering to design an ordinary car, or even a scooter, engine. Naturally there are delays, failures in meeting prescribed timeframes, and wasteful expenditures are incurred in the process. Consequently there are cost over-runs, failures, delays and eventually going back to the foreign vendor who is more likely to exploit our inadequacies. (Certain foreign experts have also identified that there are evident anomalies that suggest problems in the procurement process, which militate against the further development of indigenous capability. They quote instances (a) Why did India buy Rs 10 crores worth of parachutes from France and South Korea when the same are manufactured in Kanpur? (b) Why were 1.4 million blankets purchased from Australia in 1985 when most Indian woollen mills had export capacity? (c) Why were 1000 passive night goggles imported when the indigenous variety could have overcome the Army's objection by using a small imported component at a fraction of the cost of the whole imported item? (d) Why were 3000 pieces of illuminating ammunition for 81-mm mortars imported when the indigenous variety required only to have its fuze replaced by an imported one?" The spending spree", India Today 15 Apr.1986).

Another important reason for waste of money is a lack of proper work culture in our 'trade-union' intensive PSU's. As per the SIPRI report the cost of producing the MiG-21 in India was 193 per cent more than its imported cost. The case history of the AVRO aircraft can be very instructive in this regard. The aircraft was to be used in a number of roles in the IAF. The aircraft required a number of modifications so as to be able to meet the needs of the Airforce. Since the anticipated requirements of the IAF would not have justified the establishment of production of the aircraft within the country it was proposed that the Indian Airlines should also accept it as a replacement for some of its fleet. That Corporation, having examined the commercial aspects of the proposal, felt that this aircraft was not suitable for its needs were reluctant to place any order. In the mean while, necessary facilities for assembly were created in Kanpur with full complement of accommodation; technicians and machinery were created. The required modifications for the IAF were never fully carried out they were therefore, reluctant to place orders. The IAC was also not keen on this aircraft, even though they did place some orders after being persuaded by the Government. But at no time did the production unit have a clear prospect of sufficient orders. Consequently, leading to creation of idle capacity, unwanted inventory and a frustrated technical manpower. A production programme of this magnitude should not have been embarked upon without an economically viable order. Inspite of the best efforts of the people concerned, the ministries of Finance, Defence and communications, the Production Unit, the IAF, the IAC could not agree on certain basics.[1]

There are certain basic problem areas. These are:

(a) **The system lacks uniformity.** The user has to look to three or four separate agencies for the development and production of the same commodity. To make matters worse these agencies are usually located so far apart that any day to day communication is virtually impossible. We have no prescribed system for active interaction between those who are tasked to undertake designing and those who are to undertake manufacture and those who are to eventually use the equipment are usually not at all associated with this whole exercise. Whims and fancies of a few tend to govern the whole cycle. Even these are susceptible to change with every change in the bureaucratic structure. It may not be possible to relocate all the concerned agencies involved in a particular product in one station because of any number of reasons and considerations. However, with better communications and "internet" etc now available interaction between all concerned can be on a regular and constructive basis.

(b) **The existing control mechanism tends to break down**. To compound the problem there is often failure to coordinate fully the efforts of all such agencies. This results in a lack of unified direction and integration of development and production effort. A study of the case histories of a number of 'defence' related development projects would show that often the original factors in initiating the proposal are lost sight of. Thereafter, the project meanders along as if it were an exercise that is an end in itself. MBT Arjun a classic case highlighting our penchant for misplaced faith in our capabilities. With over twenty years and thousands of crores of rupees there is no idea as to when the troops will actually get to see the tank in actual usage.

(c) **No set procedures exist for contracts**. Different contractual procedures and procurement practices confuse industry, and tend to encourage uneconomical and often unethical practices. There are evident anomalies that suggest problems in the procurement process that militate against the further development of indigenous capability. More often than not, if the R&D and the defence industry are seen to be in the process of delivering the goods the user will find some way of thwarting such an effort. Since the contracts are designed to safe guard the interest of the state the user can always find some way to scuttle the progress if he has set his heart on a foreign product.

Note: (The Indian defence industry has been consistently retarded by a fifth column. The country's decision-makers have consistently allowed the armed forces, especially the IAF, effectively to sabotage indigenous efforts in favour of imported equipment. Over time, India could have built up a significant defence production base that would have become increasingly independent and self-supporting. Because they required Western levels of technology, projects were either bound to fail or to require increasing inputs and assistance from foreign sources)[2].

Due to the lack lustre performance of these defence production units,[3] critics now question the "sacrifices" incurred in creating a defence industrial base (DIB) in the public sector.

Both in the form of 'Ordnance Factories and PSUs when we have more pressing problems of poverty, starvation, illiteracy, ill health, and homelessness. Because of the reluctance of the government to involve trade and civil industry in products related to defence, and to sustain these monoliths the military has been, perforce, dependent on the 'Ordnance Factories' and the PSUs for most of its needs right from the colonial days. This 'forced dependence overlooks the fact that their products generally are costlier and what is worse are often substandard. And then we have legislations and official mandates that insist that certain 'state run' or so called 'welfare' establishments have to be patronised for procurement of things like blankets, towels, and other common usage items. These items are available ex trade at extremely competitive rates and with a much higher level of assurance. But rule bound bureaucracy continues to insist on such wasteful expenditure regardless of the cost and quality differential. (Certain welfare oriented small-scale units running under the aegis of either the central or state governments manufacture a wide range of low technology items. All government departments including defence are expected to patronise them. Largely these organisations are as sloppy as most of our other PSUs are. These units are given price preference over commercial bidders, fail to deliver for a variety of reasons, seek price escalation, and the cycle of non-delivery starts all over again.) It is only recently that the military has come to realise that it is not a captive market of the Ordnance Factories, or of any body else for that matter. But actually a paying customer who by virtue of the fact that it is a big spender deserves a much greater consideration from the vendor than what it has been getting so far.

This is no "Humour in this Uniform"

There are as many as 16 sizes of the uniform items. Maximum number of demands, almost always, is for extra large trousers without commensurate demand for extra large shirts. Now this trouser is 48 inches in length and has a waistline of 40 inches. It is difficult to visualise that we have very many soldiers subscribing to these specifications. One finds it to be rather intriguing. On going into details it was found that these trousers were convenient for refit, and were therefore being demanded by units with gay abandon. Quartermasters made no effort to scrutinise or verify the demand inspite of the fact that all units maintain "size rolls" so that they could project accurate data for provisioning action. It is time we adopted the commercial practice of having only three or four sizes. This will lead to a lot of savings, both to the procurer and user, because refitting etc will be considerably reduced.

Misusing Doctors

We hear of perennial shortage of doctors in our military hospitals. All kinds of allurements are offered to them to join the Service. And yet, we employ them as 'Registrars' (and certain other non medical assignments) in the hospitals. This is a purely administrative task where the charter of duties entails provision of rations, upkeep of the hospital premises, salaries, maintenance etc, of the civilian staff like ward sahayaks, safaiwalas and all that. Of course, there are the usual company commander, quartermaster and others for specific tasks, but utilizing doctors for administrative purposes is a waste of highly qualified professionals. We could easily have officers from the AOC, ASC etc, for this job. Having dealt with industrial, non-industrial and other categories of civilian employees they have the right attitude and expertise for the job. Such a step would also spare doctors for their more important and relevant profession.

There could be an argument that administrative exposure is important for an individual doctor because once he assumes command of a hospital he would require first hand knowledge of administrative tasks essential for a smooth command. This argument can not hold any ground because the level of involvement in day to day administration of the 'specialist' who gets to command a hospital is virtually impossible because of the demands of his profession. Secondly, at that level one is more of a senior manager where day to day functioning is concerned rather than a commander of troops. This wastage of doctors on non-doctoral assignments needs a look in by appropriate authority.

Army Statistical Organisation

Sometimes in the 1940's an "Army Statistical Organisation" was created to provide data on Men and Material for planning purposes. An officer of the "Indian Statistical Service" eventually headed the organisation (ISS) holding a director level rank. Later when computers came into service and a "Management Information System Organisation" (MISO) was to be created in the late seventies it was found that there was hardly any role for "ASO". The bureaucracy felt that reduction of ISS cadre post(s) was not acceptable to them because it could reduce promotional avenues where that particular service was concerned. No amount of coaxing and cajoling helped. Eventually MISO came about but Director ASO remained along with sufficient staff to justify the director's employment. The entire organisation virtually has no work but for what is created to justify the existence of ASO. This 'created wok load' has a snowballing effect on the troops because for most of the studies undertaken by this organisation military units are required to send inputs- an activity an operationally committed unit could do without.

All the three Services have similar wasteful posts either to accommodate certain cadres or created at the whims and fancies of their "Heads" at various times. In most of the cases officers who could be doing constructive work elsewhere are assigned to these tasks just to while away their time and expend 'tax-payers' money. An indirect benefit to a certain category of officers is that some more "Delhi Tenures" have become possible. Since these functionaries specifically deal with logistics at various levels there are often avoidable delays in arriving at decisions that could improve the quality of service provided to the troops. There is an urgent need to carry out an "organisational audit" and pruning of the 'higher headquarters'.

Reports and Returns

There is a 'Defence Service Regulation' which lays down the prescribed drill for introducing any report or return on a regular basis. The regulation also stipulates that if a 'return' does not have a "statistical registration" number a unit is not obliged to forward the same to the higher headquarter. Yet, every headquarters is usually busy calling for information from units that could very well be available in another branch of that establishment. Some of these returns are important for planning material support but most of them are wasteful where clerical effort, paper work and utility of these returns is concerned. For example, when an "off road state of vehicles" is called for there is no need for another report asking for details of vehicles off road for " want of batteries, tyre and tubes". Another problem is that while sending any report to Army Headquarters, or a Command headquarters a unit is expected to send a copy to the intermediate headquarters also, not withstanding the fact that agency may not have any need for that input.

At least, 30 to 40 percent of our 'reports and returns' are totally wasteful and perhaps continue to be called for in order to justify clerical establishments in various headquarters.

Transport Related Issue

It is difficult to understand the rationale behind providing "gas guzzling" 'Jongas' and 'Nissan' one ton vehicles with their 110 'horse power engines' to "peace units" located in a place like Delhi, Bombay, and other such like stations. Units in such places could very well be supplied with 'fuel efficient' vehicles available commercially. Similarly, why can a "dispatch rider" in these units not be given a "scooter" instead of 'fuel inefficient'

Bullet or any other brand of motorcycles? Besides guarding a dispatch rider against getting stranded in some far off place when he has a puncture, the scooter will consume less than half the petrol of a 'motor cycle'. More importantly, the scooter can be taken to an authorised dealer for repair and servicing, and thus saving on the engineering effort when our workshops are overloaded with all kinds of vehicles and other items needing urgent repairs.

Indian Army forbids its officers from driving 'Government' vehicles. In today's world when every officer is a reasonably competent driver it seems a waste of 'tax payer's' money to hire so many drivers for conveying officers for duty. In this one could also take inspiration from the American system where colonels & below drive vehicles themselves. They are issued with petrol & service coupons to draw fuel within the framework of authorised mileage from any petrol pump & they get the servicing done at authorised service stations. We can certainly try out such a system in metropolitan cities

A large number of load carrying vehicles are employed in taking children to schools. These vehicles are modified for the purpose and benches are fixed in such a manner that the vehicles cannot be effectively used for the basic purpose for which these are meant. The vehicles also have a limited passenger capacity, thereby creating need for modification of still more load-carriers. If we provide proper buses for the purpose, which besides making journey of the school-going children more convenient would also provide a much larger number of seats, thus reducing the number of vehicles required for the purpose. Savings in terms of fuel consumption, employment of drivers and speedier movement of stores due to release of load carriers from school duties will pay off the expenditure incurred in procuring buses. That, these buses will come in handy when any troop movement is to take place will be an indirect benefit to the system should be an added incentive for taking this money-saving decision.

Some Solutions to the Problem of Waste

There have been any number of studies undertaken by various organisations and institutions to identify wasteful areas in our 'material support system". However, due to considerations other than merit of the case the findings of these studies have only been partially and insignificantly implemented. This failure to do so has been a major cause for our having lost out on opportunities for bringing in significant improvements in our logistics functions. A few ideas are appended herein.

First, bring in the concept of "Total Cost Management" (TCM)[4] where our logistics are concerned. Every activity needs to be costed and cheaper and more efficient methods are required to be adopted. For example, in our Central Ordnance Depot at Chheoki a total of approximately three truckloads of stores are issued every day. Assuming that one third of the manpower (about 700 labourers with an average salary of Rs. 5000 per head) is involved in the issue activity the cost of loading one truck comes to

$$\frac{700 \times 5000}{3 \times 25 \text{(working days in a month)}} \; 3 = \text{(approximately) Rs. } 43000$$

Reserve Supply Depot at Lucknow hires contract labour for the purpose on 'as required' basis and they pay Rs. 600 only per vehicle to the contractor. It might not be a bad idea if our Ordnance, ASC and Engineers' depots also cut down on their permanent establishments retaining only 40 to 50 percent of workforce and contracting out the rest. Certain Naval depots and dockyards are following this practice where the requirement of work force is not of perennial nature. In the case of our Army's depots there could be an immediate gain in the form of other savings as there will be a commensurate reduction of 'clerical' establishments and improvement in the performance of the residual workforce.

Next, DGOF and PSUs involved in defence production are in need of immediate restructure and reorganisation.

They must become economically viable and competitive. They should shed surplus man power, close down those involved in low technology items like clothing and general stores, go in for more and more 'outsourcing'. The factories retained thereafter be made autonomous and must reduce non-essential over heads and focus only on the core elements of their field of manufacture and supply to the military. Accountability of the management should be direct to the customer so that indiscriminate price escalations, cost over runs can be cut down.

Notes and References

1. *Defence without Drift*, by P. V. R. Rao.
2. *India's Adhoc Arsenal*, by Chris Smith.
3. Maruti Udyog Ltd. achieves an output of 24 vehicles per man per year. The output achieved by Vehicle Factory, Jabalpur, is 0.6 vehicles per man per year.
4. It covers all functions involved in making support available to the combat troops. And if wielded wisely, acts as a crucial bridge between strategy and operational efficiency. TCM translates itself into a universal set of all attempts to improve functional efficiencies in a way that lowers the cost of carrying out a specific function.

13

LOOKING AHEAD

Operational plans, contingency plans, operational orders, SOPs etc, are no more than a framework within which operations take place. Ideally, if every one is fully briefed and in full picture of the whole thing and if the events go strictly according to plan, then an operation should be able to take place without further orders and instructions. Often enough this does not happen. Too many unpredictable situations, unforeseen events and occurrences overtake the most well thought out plans where these factors concern logistics back up for the operational contingencies. Earlier, life was simple for the commander as well as the lowly foot soldier. A sword, a horse and may be some ration were all the logistics needed for launching a successful campaign. From the Stone Age and smoke signals to thermonuclear devices and satellite communications the world has come a long way. With the growth of science and technology the society changes. When that happens every thing about the military also changes. Not only its technicalities, the tactics and field doctrines but also the standards and methods of training, levels of its professional knowledge, the means of functioning and the modes of equipping and maintaining the forces also undergo alterations. The exponential rate of development of military technology and adoption of what primarily were civil end usage products the military world has come to a potent stage. Now an army if not moving fast enough to absorb the technology, could be left standing behind in the race for superiority in its functional efficiency, and the country in its industrial and economic growth. During the last fifty years the environment in which our Logistics systems operate has altered radically. But our management of resources has grown only incrementally without a comprehensive look being given to the subject in its totality. Resultantly, patchwork efforts to fall in line with the environmental changes have not allowed us to reap full benefits of the investments made. Over time importance of logistics support (at less cost) will grow as the economic challenges and threats of this "New World" environment becomes more prominent and competitive. This means that India will have to spend more on the day to day operation and maintenance of its forces and equipment simply to "stand still" as costs of the equipment goes up and technology makes it more expensive. At the risk of an understatement, it can be said that national security and defence are indispensable goals in the plans and processes of nation and state building. Without this the plans for development, social uplift etc, are public relation exercises which may or may not work. Often these do not work we learnt the hard way in 1962. With resources becoming scarce and with defence equipment and its maintenance becoming costlier a responsible government has to draw a balance that meets aspirations of the people alongside a strong defence. For this, realism, or at least prioritisation, will have to be introduced both to the defence budget, how it is spent where equipping our forces is concerned and to the cause of economic upliftment of the people.

Universally, the thrust is to apply thoughts and brains to the expansion of the power of material and personnel without increasing either. At the national level, all over the world, individual countries have taken, and are taking steps to increase logistics' efficiency with a view to reducing unnecessary expenditure on the military tail. We have been rather slow in accepting the fact that ever changing and growing military technology has its impact on the fighting capabilities of our military. With the equipment and its upkeep becoming more and more expensive the nation cannot afford to have its resources being spent in a manner that does not become cost effective. There is a "Financial Planning Directorate". Its mandate, however, does not include "Total Cost Management"(TCM), a key element in any professionally run organisation. True, to an Army rooted in the past, let alone prepared for the future where financial constraints do not permit modernisation of our depots etc., a question of total cost management may sound to be heretical and revolutionary. There are too many hidden costs in our system.

What is TCM?

TCM translates itself into a universal set of all efforts to improve operational efficiency in a way that lowers the cost of those operations. As a 'thumb rule' every activity in the supply chain must aggregate to offer the end user an ability to carry out his operational tasks at a cost that is vastly lower than the 'value' it generates for him. That means there must be complete cost-control along the entire value chain, and not just in individual activities. Central to TCM is the principle that the entire value chain be made faster, better in its operations and more responsive to the troops at reduced costs. TCM is concentrated on maximising productivity by ensuring that the right person is employed on every task. In our depots, workshops and production establishments TCM demands elimination of wastes and delays. (Certain industrial houses have costed the effort involved in issuing and accounting for "meager value" items. Where the price per piece is less than the cost of clerical effort, paper work, policing etc., the items are left unattended in their respective bins and any worker can take what he requires without making any indent for the same. In our Army the indenting, controlling, accounting and issuing procedure, including audit by the 'local audit officer' for every item is the same regardless of the cost and its availability state. If we work out the cost of the entire process and modify our procedures a lot of money and effort could be saved. We could surely learn from the industry and redefine our systems where we bring in a little trust in the users of materials and the custodians of the government property.) In Research and Development establishments TCM should manifest itself as the objective of designing a product for which manufacturing is a low cost process. Not only that we also have to ensure that when particular equipment lands in the hands of troops its maintenance and upkeep cost is not unnecessarily high. The cost management culture along with enhanced material support should be our 'mantra'. Then only our 'service units' can become more responsive to the front line soldier.

There have been a large number of studies, directives and instructions to economise. There has, however, not been any specific effort to formulate a strategy that allows us to draw more mileage out of our defence budget. At best all that has been happening is that some conventional cost-cutting measures get adopted from time to time and after a while for reasons other than merit of the case the system gets back into its familiar groove. One cannot emphasise too strongly that the management of logistics is a pro-active task and the way to minimise costs in this area is not following a do-nothing, spend no-thing policy. Logistics is a critical component for ensuring operational ready-ness and effectiveness of our troops. There is only one way to provide a better service to the troops- continuous effort to manage costs and make strategic investments where and when necessary. Ordnance alone holds an inventory priced over Rs. 40,000 crores, occupying considerable real estate and employing a huge work force consisting of both combatants and civilians. We have a massive 'stores budget', we have huge infrastructure for maintenance of equipment but we have no infrastructure to monitor the costs involved in providing material support to our troops. Drawbacks are many. Ill maintained and dilapidated storehouses, lacking in visibility of stores and labour intensive work environment lead to large inventory and excessive drain on Army's budget without commensurate value for money thus spent. In our system it takes years and years for the government to make up its mind. More often than not, by the time a decision gets taken the technology itself might have advanced further. One can say that the delays allow us to go for the later products, and thus may actually be saving money, not realising that it takes considerable time and effort to absorb technological changes. Unless the transition is by stages, chances are that there will be a mismatch in our training, revising of battle procedures, grooming of the repair and maintenance back up, and even management of inventory. It will be good to remember that ever evolving armament technology and consequent refinement in military thinking creates its own problem of finding both men and material that could match up to the latest in the field. New and modified equipment requires highly trained maintenance personnel to keep units combat ready. Our systems have to undergo radical changes if we want to project ourselves as a modern, well groomed and a well equipped military. That involves numerous facets of our resources and efficient deployment of these resources. Induction of newer and latest technologies in different fields will be the first step in this direction.

Any military that continues to operate with a logistics system designed for one environment, but operating in another, is bound to incur economic inefficiencies and operational ineffectiveness. In this era when army is dealing more and more with insurgencies and counterinsurgency operations and needing more of our resources the logistics operations supporting these activities are those designed when armies over the world felt it necessary to take and hold large land masses. The systems have been altered only marginally. In that, while the physical aspect of replenishment has been changed in some instances, the approach to the requirement has not been changed. It takes courage for the top military leadership to question a basic logistics support philosophy which has (generally) withstood the test of time, and which they have now inherited to administer. This is particularly so when that traditional system has been guided by the emotion "nothing but the best for the fighting man". Financial constraints, lack of 'in house' expertise, reluctance of potential foreign suppliers etc force a compromise on the military leadership. They are then forced to turn their thoughts to ensuring an adequate effectiveness for the money and men available. It is true that getting people to switch to a low cost high value paradigm is not going to be easy. It is always difficult to persuade busy commanders, politicians and bureaucrats that there might be newer and better ways of achieving the same or even better results. That they are skeptical of good ideas until they take a practical shape is easily understandable. There is, however, no choice but to force cost-consciousness to pervade every activity connected with material support to our troops.

But, before starting the ground work on a technology based logistics system where total cost management is the guiding factor a lot of thought is required to be given to the ultimate design of such a system. Any policy or planning we adopt will involve a material support strategy to shape the future in a preferred direction. It must also encompass assessment about the future and its impact on the Logistics Service's objectives and goals and how do they fit into the Army's deployment and employment doctrines. This necessitates a knowledgeable and data based conceptualisation of future trends within the system and outside it. In order to develop something, which could go a long way and should withstand anticipated changes in the logistics support systems based on changing operational doctrines and improved armament and communication systems, a detailed assessment of likely future requirements and expectations from the logistics systems has to be made. Such an assessment has to be based on careful evaluation of trends in different areas relevant to the goals and objectives to be achieved. It cannot and should not be just a linear extrapolation of current trends. Therefore, examine your future strategy, in order to incorporate futuristic planning in such a way that the present systems which have by and large stood the test of time, could be weaned away from, without any dislocation of the service that we are expected to provide to the troops. This means a much higher level of responsiveness by the logistics units in a given situation. While our logistics posture is not at the high level we would like it to be, there is no cause for painting too grim a picture. We have a reasonably efficient communication system. We have a defence industrial base; we have technological and industrial infrastructure that is amongst the leading in the world. And services like medical, repair facility, research and development both in the private and public sector measuring up to international standards have come up over the years. All that we need to do is give a new look to our logistical support to the military if we want efficiency and reliability without increasing our defence expenditure.

Discussed hereafter are some of the steps we could take to bring in better responsiveness and improve quality of service provided to the combat troops.

14

Automation and Logistics Management

"Knowledge", they said, "is power". One such powerful medium that provides the means for accurate and fast information, and thus knowledge, is the computer. The dramatic advances over the last 30 years or so in information creation, collection and dissemination are taking down the boundaries of distance, time and space. Information, and therefore knowledge, is one of the most important commodity decision-makers in all walks of life use today. In our environment, a large army has to be clothed, equipped and sustained over a vast area with different climatic, atmospheric and operational commitments. The traditional means of obtaining information on logistics and taking decisions based on that information are no longer sufficient to meet the exploding need for right, accurate, timely and more detailed inputs for providing an efficient and reliable material support. The requirement for these inputs assumes special importance when we see that the entire process of procurement and holding of inventories is fraught with (literally) hundreds of uncertainties both internal and external to the organisation. Internally, uncertainty exists-with the top management whose decisions affect procurement. The 'file bound' system that we have never allows a material manager to be certain of his bids for procurements finding approval at what level and stage of decision making. Logisticians have to deal with unreliable suppliers, unpredictable demands, long and unspecified lead times, failures in supply and rejections due to poor quality. Users, who indulge in hoarding, inflate their demands mainly due to lack of confidence in the logistician's ability to meet every demand add to the problem. Externally, uncertainty exists in the form of government regulations, quotas, licenses, bureaucratic unconcern, availability of materials within and outside the country, suppliers who are beset with their own internal uncertainties; the technological infrastructure of the country; labour unrest; railway strikes; natural calamities; international relations- the list, in fact, is endless. We cannot control or check most of these factors. But we can, based on experience and knowledge, dilute their impact in so far as operational readiness of our troops is concerned. With the technology evolving at such a brisk pace it is envisaged that tomorrow's logisticians will operate in an environment where reaction time is short, assets are scarce and too widely spread out in different types of terrain. And, as usual there will always be a paucity of funds for meeting all the demands placed on the system. With an inventory of over 5 lac items that the army holds and with a troop strength of over one million deployed through the length and breadth of the country it has never been an easy task to effectively and economically monitor the whole activity while ensuring a reliable material support. Measured by any yardstick, the immensity of the task entrusted to the Services in terms of diversity of range and increasingly sophisticated equipment at the wholesale and retail levels over widely dispersed locations is challenging.

Introduction of automation in various other walks of life has made us conscious of the fact that in so far as equipping and sustaining the forces is concerned, a very powerful tool had come into existence. Modern technology has permitted the introduction of computers into logistics units. Later day communications have made it possible to pass the demands automatically through the whole chain of command right down to the base depots. This enormously improves management capabilities and increases the speed of supply.

Computerised stock control can make the handling of demands far more fast than ever before and can present the commanders with an immediate 'read out' of holdings and reserves at various levels. This information can be flashed onto the 'computer monitors' simultaneously at all levels of command. This will, in turn, help the operational commanders to correctly assess their requirements and supply positions in so far as their immediate needs are concerned. The supply position of "light-weight" vital components in respect of tanks, artillery guns, communication equipment etc can be predicted more accurately. Thus, automation provides us with better and more reliable inputs for taking material management related decisions both in peace and war. This realisation has been registered at various levels and action is in hand to design and develop logistics related "soft wares".

Automation and Resource Management

Logistics management is becoming increasingly complicated with armament technology and troops deployment getting to be more and more dynamic. It becomes all the more difficult to keep trace of the assets when the troops move from one area of deployment to another, often leaving their main equipment behind. They generally carry the spares and accessories along where, in all probability, they wouldn't need them in the foreseeable future. The replacing troops who come to hold and use this equipment express their need for those spares and accessories, items that often become untraceable for a fairly long time. This adds that extra bit of uncertainty while working out the provision figures for procurement because the greatest single factor which determines the quality of logistics support is the efficacy of requirement determination, procurement and correct distribution against the back drop of progressive modernisation of the Army. Doing all these tasks manually with the kind of inventory that we hold is always susceptible to inaccuracies. With the power of the computer to quickly and accurately react to information needs regarding availability, dues in, location and in-lieu usage of stores and the origin of a demand ;it is possible that certain inherent and latent problems in providing a more efficient logistics cover will be resolved. Need for a system that provides means of automated data processing and high-speed digital data transmission so that an economical and efficient allocation of resources could be made, was never greater than what it is now. Automation of numerous activities connected with the management of these inventories will give a more accurate and timely estimate of our requirements. An accurate and speedy computer- based "resource management" system has become essential today for efficient management of material support and for giving confidence to the supported troops in their logistics back up.

Activities to be automated

There are any number of essentials (within the ambit of the term logistics) which, when accurate and timely data is available, can materially improve the service to the combat troops. However, for ease of explaining the hypothesis and because author's experience, in so far as designing computer based systems is concerned, is primarily confined to "Ordnance" inventory, the discussion on the subject is related to the Ordnance aspect of the logistics support. With some modifications and changes commensurate with the 'role and tasks' of a particular Service these parameters can be applied (and if required expanded) in any instance.

Inventory and material support to the troops involves a very wide and diverse range of activities. It also entails interaction between various agencies within an organisation and outside. No inventory or a dead inventory leads to failure of the system or at the very least, loss of organisational credibility because a failure to make material available to combat troops when required by them is an unacceptable situation to any commander. This means a very detailed, careful, and practical analysis of the requirement. The size of inventory, its dispersion through out the country where troops are deployed on all kinds of duties and our financial limitations demand that a tremendous amount of care and foresight is used when working out the requirement figures. The whole exercise in itself is a time consuming activity. The depth and the range of inventory held also dictate simultaneous and concurrent activities at various levels within the organisation. Broadly these activities are:

(a) **Purchasing** - The purchasing responsibility has often been defined as "buying the right material, of the right quality, in the right quantity, at the right time, at the right price and from the right source". The purchasing action, in itself, involves a large number of activities which take place at various levels of 'resource management. First and foremost is 'forecasting'. It is defined as the extrapolation of the past into the future. It is usually an objective computation involving data, as opposed to prediction, which is a subjective estimate. Accuracy of the extrapolation depends on conditions holding in the future, particularly the aging of the machinery, armament, vehicles, and communication systems. The plans for upgradation and frequency of maintenance will also provide an important input for the purpose of forecasting. In the case of weapon systems, armoured fighting vehicles and communication systems, the residual life span ie the likely duration in terms of years or months (the period) for which maintenance is required to be planned and provisioned for will dictate the parameters. Again, this factor has a tendency to be erratic because the replacement policy or schedule may be changed for a variety of reasons. It could happen that the forecaster had not made a fresh procurement bid because the stocks in hand and in the pipeline had been found to be adequate for the predicted residual life span. In the interim a policy change enhances the period for which maintenance is desired. With the long and unpredictable lead-time of our defence industry and uncertainty where the equipment happens to have been imported could always put any forecast out of sync-computer or no computer.

One may be able to forecast more accurately in those cases where the demand is mainly determined by maintenance activities of a repetitive nature or where the equipment has been in active usage for a considerable length of time and a pattern has been established. For a resource manager, any of the following or a combination of more than one of the following types of analyses, will be required to be undertaken:

- **ABC analysis** - This is related to the annual usage cost of a particular item. 'A' items cost approximately 60 to 70 % of the budget, and they are lesser in number. 'B' items cost 20 to 30 % of funds, 'C' items are always in a larger number but cost hardly 10% of the money available.
- **VED Analysis**- This relates to vital, essential and desirable status of spare parts, clothing items, rations, medicines and so on. As the terms imply, certain parts and assemblies are considered to be vital for meeting operational requirements. Similarly, some ration and clothing items have been identified as 'vital' in so far as high altitude operations are concerned. In the case of 'life saving drugs' and where all those are located can make important difference. While making a forecast these aspects are always kept in the perspective.
- **SAP Analysis** -'Scarce, available, plenty' allows us to build into the provision forecast the 'Availability State' of an item. Quantity to be ordered will also be governed by the 'scarcity' factor. If it is imported equipment then the likelihood of the supplier

changing his production line or the model will be another guideline in deciding the procurement policy.

- **FSN Analysis**- 'Fast, slow and normal', determine the consumption pattern of an item. However, peacetime consumption, when considerable equipment is 'mothballed', may not give us a realistic estimate for taking provision action with regard to spares and assemblies. This could cause an imbalance when operational needs are not accurately worked out in the absence of data. In a computer based system a " simulation" model can be developed for this purpose. The manufacturer can also provide some input for achieving accuracy in predictions for the future. One thing to be guarded against is that the foreign suppliers, however, may give an exaggerated database just to palm off their surplus inventory.
- For the unforeseen, other factors like 'Store Margin', Reserves etc., are built into the forecast. These are normally based on the 'general staff' indications; experience with regard to potential supplier's capability to adhere to delivery schedules. Certain other requirements, which though unpredicted, are likely to arise during a 'provision cycle' are also taken into account. However, accurate predictions on troops deployment on IS duties, UN commitments etc., are often difficult to make. Therefore, even in a computer-based system chances of inabilities arising due to certain unpredictable data cannot be ruled out. The difficulty can be partially overcome by making use of historical records built into the automated system of provisioning.

(b) **Inventory Management**- classically, defined as "the maintenance, upkeep and assurance of adequate supply of something in order to meet an expected pattern of distribution of demand for a given financial investment". In Army parlance it would mean creating, holding, preserving, and making available all the material requirements of the dependent troops in an economical and cost effective manner. Activities involved are almost akin to those in industry, excep that in the instant case failure to measure up to an operational urgency could have serious consequences. Environment for holding and preserving inventory items may not be ideal. This means factors like "dispersion", "turn-around time", terrain where stores are to be held, and capacity of a particular location, chances of enemy over-running the location etc will have to be kept in mind while designing 'inventory management system'.

(c) **Warehouse Management-** All the time an inventory manager is concerned with the task of ensuring that the troops get the right item in fastest possible time frame. In a manual system it may not be possible to manage all the activities- like controlling of the demand, identifying and selecting the right item, packaging as per the likely mode of transportation etc, certainly not concurrently. With the help of a computer he could carry out the tasks speedily and more efficiently. Faulty and incorrect identification, particularly in the case of newly inducted equipment has often created problems of increasing the 'down time' of the main equipment. 'Bar coded' inventory reduces risk of such a mistake being made when the concerned staff is not very familiar with the components of a weapon system. In a computer based system one can create a file of drawings of all the assemblies, sub assemblies, and even each minor component. Therefore, a store keeper, or a technician who only has a small bit of information regarding the item required can identify his specific requirement accurately because any "key" input will trigger the action sequence for the user to correctly identify the item required. More often than, not stocks are dispersed in various regional depots for reasons like potential area of specific equipment holder's deployment being more than one or means of communication in a particular theater being uncertain etc. In a computer-based system such stocks create greater availability, as asset placement in a global template[1] is known at all decision-making levels. This ensures flexibility in employment of the equipment and manpower besides reducing the quantity required to be otherwise, held in various locations. Identification of what a particular item can be used in lieu of or what can be used in lieu of a particular part improves over all level of assurance besides providing flexibility to the inventory manager. This also helps in preventing waste because concurrent usage of spare parts is possible even if a certain type of equipment goes out of service. (Cross checking of Issue in lieu is a tremendous advantage both for the material manager and the medical officers who can easily verify and identify the alternatives available in a critical situation).

In a computer-based system in the "Client Server" environment with all the work-stations connected to a 'server' the review can be carried out in one go because any inventory-related activity will trigger.

- Records at numerous workstations. Accurate provisioning action can be taken if all the relevant inputs have been correctly fed. Secondly, more efficient method of 'Review' incorporating Analysis indicated earlier can be made use of.
- Automated demand / supply order printing can be ensured by designing the system in such a manner that the moment a re-order level is reached the system itself takes action to refill. The time taken in generating demands and informing all concerned can be considerably reduced, as the central server will update data at all 'work stations' informing

every body of the action that has been taken to ensure replenishment.

- Monitoring of Budget is another important activity that is much more realistic, accurate and timely in an automated environment. Higher management can effectively monitor the expenditure pattern in relation to allocations. At the Working level a watch can be kept on price, payments and booking of expenditure under appropriate heads.
- Another important aspect is that of 'Audit'. This is a statutory requirement and cannot be wished away. It will be appropriate to incorporate an "Audit Representative" in the team so that the system at the design stage itself incorporates the mandatory audit aspects.

Note: Normally there is a tendency to automate existing manual systems. These manual systems have withstood the test of time. These have been found to be perfect provided one has made use of them in an ideal situation. Since neither the users are ideal, or the times, it would be prudent to think afresh while automating. Retain only those operative, statutory and legal requirements of the manual system that are obligatory. Then only one can make full use of technology.

Lead Time Management

If the 'materials manager' can accurately predict the availability then considerable savings can be made because the dues in can be staggered, the payments can be timed with supplies and the user can be given fairly accurate estimate of the time frame in which to expect supplies. For this the resource manager needs to monitor both the "administrative and technical lead times. The module for this purpose should normally include the following:

Administrative Lead Time

(Defines the time taken from identification of a need to allotment of funds and placement of order on the supplier.)

Main inputs are: When the need for an item was identified, ie, when the "annual provision review" or any special review was carried out leading to arriving at requirement of a particular item. When the requirement was projected to the appropriate financial authority. What has been the flow of indent, ie, has it followed correct channel or sent to destinations which have no role to play in procurement. Where required the higher authority whose intervention could move matters can also be identified.

Before placing order a vendor analysis would have also been carried out based on the past performance and his capability to deliver goods well in time. Weightage can be given to those factors that identify a particular and specialist class of vendors only. This will help in arriving at a more realistic and accurate details while considering placement of orders or staggering receipts.

Technical Lead Time

- Workout the schedule where the 'possible date of supply' is clearly given. It should be realistic and practicable and based on actual past experiences with the vendor. Identify the stage at which the vendor is required to be expedited
- Depending on the value and criticality of an item the level of monitoring be identified. The system should throw up relevant details for action by the nominated authority.
- Keep track of the supply schedules and if a transportation model has been made operative then the system should highlight destination wise delivery.
- Effects of slippage in delivery schedule should also be highlighted so that the higher management is kept informed. If required, that authority can take suitable remedial action well in time.
- Does a slippage lead to any operational set back, price escalation etc. And if so, what is the alternative

available be identified. This factor will be of particular importance where 'vital' items are involved, or where the user is located too far away from the "supplying depot".

While taking decisions on the subject of material support after examining the above factors we should safeguard against obvious pitfalls of current management theories because in the case of a shortage ridden army like ours there will always be the temptation to invoke the "industry pattern" of resource management. There is a general (and perhaps misplaced) awareness that several of the key factors that private companies use to gain a competitive edge and greater profits are considered to be important to the effectiveness of many military organisations, especially those with combat service support missions. In various discussions about material support one has been hearing of 'just in time' inventory management, 'total quality management', 'theory of constraints' and what not. In civil industry these competitive-edge factors include product and service quality, system efficiency, on time delivery, and customer service. It is laudable to aim for all of these. A major pitfall however, is the hawking of these 'newly fancied' management jargons or wanting to apply those resource management techniques which commercial establishments adopt to enhance their profitability to the mission of providing logistics support to the military. The advocates of these theories seemingly miss the point that these theories are designed for profit-making organisations; their objective is to increase a company's profits by improving its competitive edge and reducing its costs. In the case of a profit-oriented organisation if a particular assembly or a spare part that is vital for the production cycle does not fetch up, the worst that will happen is that the plant may have to close down for a while. Now imagine the scenario where the Army has adopted the JIT and a part or a major assembly that is programmed to arrive at a particular time and location does not arrive. This could put a sensitive communication or an important weapon system out of action at a wrong time, leading to classic situation of "for want of a nail." Can the military afford this luxury? Many of these procedures and techniques cannot be applied to military organisations without some modifications. This is so because we function on a basis that is different from commercial enterprises; we are not motivated by the need to earn profits, and improvements of debatable value at the expense of reduced effectiveness. We, as managers of costly materials held by the military, should broaden our focus beyond the specific procedures and techniques to understand the logic underlying these new management systems.

Having examined the facets that have to be included in an automated system of 'resource management we should go in for a discussion on the essential ingredients which make a support system efficient. Ease of Operation will be our first requirement. There is a general apprehension that the computer is too technical for an averagely educated person. This is not so. A user has only to feed in specific commands for the machine to provide the required output. That does not require "computer literacy" other than "press any key". System should be robust enough to withstand vagaries of its usage. Be it at the base depots or forward echelons of logistics support installations. All those handling various facets of the system may not be proficient in doing this task. Due to some wrong procedural command or due to a mix up in activating certain functions it can happen that the "confused state" leads to a 'break down'. To safeguard against such a situation, the system should be able to "roll back" to its meaningful state, and be ready to be responsive with more proficient handling. The system should be capable of functioning without any specified or unspecified problems at any stage. It should have a capability to absorb those changes that could come up even later when the user, having become wiser, wants to incorporate certain new aspects..

- **System Software**: It is not unusual for the "soft ware industry" to come up with newer versions having some additional features so as to ensure its market share. To guard against their 'hard sell' it would be advisable to select that software which meets the predefined and identified requirements. It should have built in resilience to take up fresh additions, which will always come up once the show gets going. With the help of 'open data base connectivity' (ODBC) it is now possible to use any software in conjunction with another. In order to avoid getting saddled with one vendor and/or type of hardware the software should be hardware independent.

- **System Hardware**: The computer industry is changing so fast that one could have floated tender for particular specifications and by the time the tenders are received and opened a better and more powerful version could already have come in the market. However, with the power of computer going up the prices almost always come down. Insist on the vendor supplying the latest hardware available on the date of implementation. The vendor will not be a loser if he supplies the latest equipment. It will, however, be pertinent to mention that the 'user requirement' be pegged at a particular level based on the actual needs. Otherwise one would never be able to complete the task if efforts are made to include every new feature whether it is required or not in the hope that may be a need comes up in the future. Every Service has its identified and pre-determined tasks, certain modifications to the existing procedures and systems may come up as an afterthought, requiring a change in the system configuration. This can be an add-on. However, a well thought out "Requirement Analysis and Problem Definition" can cover most of the loose ends.
- **Security of Information**: One price that is to be paid when we use computers is that an expert can almost always break codes and obtain access to the data. Therefore, when designing a system care should be taken to build in sufficient means of security.

Depending on the sensitivity of information safe guards can be built in. The American Army is already using certain terms to define the level of secrecy required to be built into a system. We could make use of these parameters, as these are freely available in the market.

Spin offs of Automation

First and foremost is savings, both in terms of money and resources. With the 'provision figures' being much more realistic and efficient management of inventory in hand considerable savings (in financial terms) can be made. With the use of computers it will be possible to quickly identify the spare parts required and reach them to the user. With an improved support the user will gather greater confidence in his logisticians, leading to a situation where inflated demands will reduce, hoarding will decline and greater number of troops will get a much better service. With obsolescence becoming better predicted one can guard against waste which normally surfaces when the equipment goes out of service and one finds that either the unit or maintaining agency is holding too many spares. More important will be the fact that the issuing agency can monitor demanding pattern and identify excess usage, and where necessary, take remedial measures. It is not possible to do so in a manual system because the number of demands received in our depots are far too many for a detailed scrutiny.

On line MIS for Efficient Management - Information is most important for decision making, particularly with regard to placement of orders on the suppliers, monitoring of receipts, and physical movement of stores. The System can be directed towards a limited set of well defined, predetermined and well understood applications. Higher management can make use of Information thus supplied for making allocations when the inventory holdings are not sufficient to meet all the demands placed by the dependent troops.

- **Elimination of Obsolete Inventory**: Obsolete inventory is a dual loss if we continue to hold it. Firstly, it occupies space required for items in current usage; secondly, unless disposed speedily it may not fetch appropriate price on disposal. Difference between developed, developing and under developed countries is that, almost, always equipment being phased out at one end is under introduction at a lower level. One should be able to locate a potential user in a sister service or 'friendly' foreign user who might be a step behind in so far as upgradation is concerned. The system should quickly identify "in lieu" usage etc and list out the items which will have no future utility in so far as a specific main equipment is concerned. This is particularly necessary for the 'East European' equipment because their degree of standardisation is relatively higher and they make effort to economise. It took us some time to find out that the "Schilka" gun is mounted on a PT76 tank whose spares we were holding in plenty at the time of phasing that tank out of our service. If this information was available earlier we could have saved on purchase of spare assemblies and subsystems in respect of the gun.
- **Better Utilisation of Manpower**: If certain spare parts are not available, or there is delay in locating and supplying them to the user trained technical manpower in our workshops and laboratories could be under employed. Computer based system will cut out a large number of functional delays and thus may keep the tradesmen appropriately busy at all levels. This will also ensure better exploitation of the equipment in the hands of the troops. In the absence of a realistic and efficient system it is not unusual for the "warranty period" to expire awaiting spares in case of an unforeseen or unpredicted break down.
- **Real Time MIS for 'High Cost and Usage' Items**: It has been experienced that not more than 5000 items out

of a total inventory of over 5 Lac are critically important, and cost almost 60 to 70 percent of the total annual budget. In a computer based system one can ensure that the availability of these items at various echelons and the locations is known at any given point of time. The higher commander can be kept informed about the stores in pipeline so that the operational plans and the troop's deployment can be made accordingly. Since a realistic and accurate data covering all aspects will be available, one can stagger his receipts, plan consumption of the allotted budget, and can even adopt a transportation model where the items are heavy, sensitive and high-priced.

There is always a common grouse that material involves too much of paper work, that most of it is irrelevant and wasteful. Therefore, when we make computer based strategic plans focus should also be on improving the procedures and processes. Where the current processes are either less than optimal or have potential for stream lining using the latest technology such as open-system architecture, inter-connectivity environments, electronic data inter change, intelligence etc., the necessary changes be brought about while designing a 'computer based logistics management system'. Participation of the users is essential to achieving this programme's objectives through out the material life cycle, from the conceptual stages to the development of functional requirements, and on through operating capability because the whole aim of the exercise is to raise the satisfaction level of the 'supported troops'. The designing of a system, for it to be most effective, should normally follow the following sequence. Based on the "Domain Knowledge" and other factors peculiar to a particular Service appropriate changes, both in the sequence of action and implementation pattern can be made:

In the ultimate analysis making material available to the troops at the desired location and time is what material support is all about. With the usage of computers in logistics

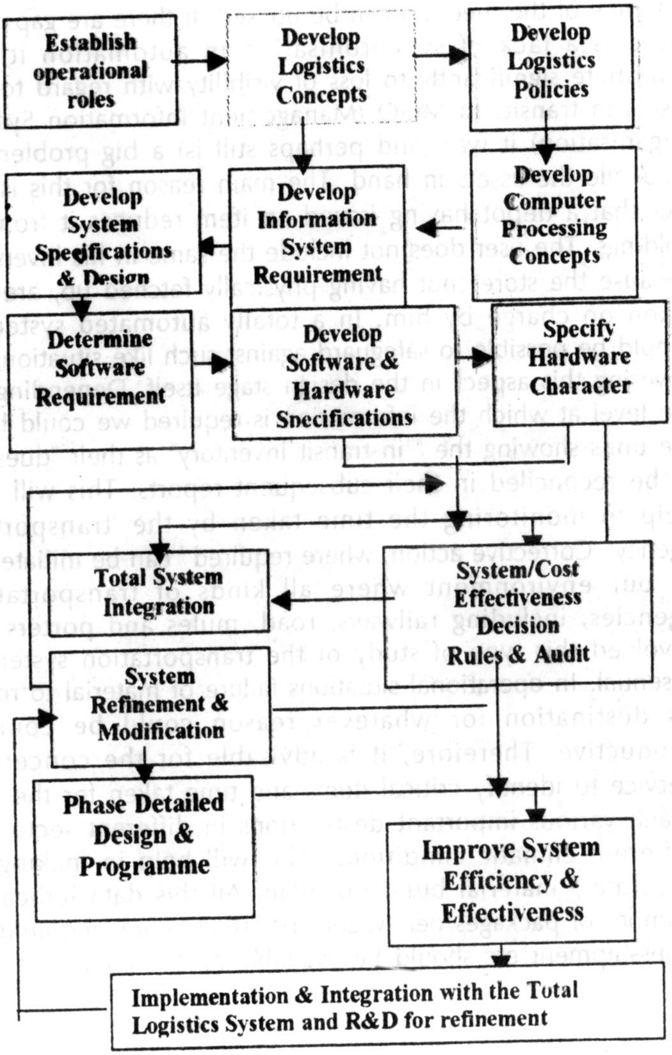

Establish operational roles
Develop Logistics Concepts
Develop Logistics Policies
Develop System Specifications & Design
Develop Information System Requirement
Develop Computer Processing Concepts
Determine Software Requirement
Develop Software & Hardware Specifications
Specify Hardware Character
Total System Integration
System/Cost Effectiveness Decision Rules & Audit
System Refinement & Modification
Phase Detailed Design & Programme
Improve System Efficiency & Effectiveness
Implementation & Integration with the Total Logistics System and R&D for refinement

management it will be possible to provide relevant inputs for an efficient and effective material support system. By and large a Service would know as to what exactly are its problem areas that need rectification and corrective action in order to make the service more responsive to the troops. All these factors can be built into "Requirement Definition and Problem Analysis. Automation has to be a well synchronised exercise at all levels of "resource management". Then only full potential of the medium can be utilised. If there are gaps and there is a lack of synchronisation in automation it can contribute significantly to loss of visibility with regard to the assets in transit. In MISO (Management Information System Organisation) it was (and perhaps still is) a big problem to reconcile the assets in hand. The main reason for this is the fact that a depot having issued an item reduces it from its holdings. The user does not include the same in his inventory because the stores not having physically fetched up, are not taken on charge by him. In a totally automated system it should be possible to safeguard against such like situations by covering this aspect in the design stage itself. Depending on the level at which the information is required we could have the units showing the " in transit inventory" as their "dues in" to be reconciled in their subsequent reports. This will also help in monitoring the time taken by the 'transporting agency'. Corrective action, where required ' can be initiated.

In our environment where all kinds of transportation agencies, including railways, road, mules and porters are involved this type of study of the transportation system is essential. In operational situations failure of material to reach its destination for whatever reason could be counter productive. Therefore, it is advisable for the concerned Service to identify critical items and time taken for them to reach various important destinations in different sectors in different climatic conditions. This will help in making an accurate " material build up" plan. All this data indicating number of packages per wagon, per truck, effort required for transshipment etc should be available on the computer. This will help in identifying correct medium for transportation of material.

Bar Coding

Automation of the inventory control and other logistics functions where bar coding is also included could really be a big help. Today, it is not uncommon for a demand either not being met or wrong spare part supplied due to faulty identification, particularly in the case of a newly introduced equipment and specifically in the case of communication/ signal inventory. In peacetime such a mishap can, perhaps, be absorbed but in a critical situation the consequences could be catastrophic. This usually happens because the storekeeping staff in our 'base depots' being civilians are not tuned to undertaking identification of newly introduced items, nor are they even trained to do so. The system advocates "on the job" training only where a participant normally gets to be aware of specific items and their application to a particular weapon or communication system with the passage of time, when quite a few issues have been made. This inadequacy of the material support system manifests itself when sometimes the user units have to send their representatives to our 'depots' and carry out the identification of the "demanded" spares and accessories. This situation can be largely overcome by using "bar codes".

Since the logistic resources are always going to be scarce in any future conflict, by bar coding and asset tracking we could make the best use of those resources. In the US Army they have introduced the system where a simple 'button' size device is located in the shirt collar of every soldier who is operationally committed. All necessary details are endorsed on this device. If the man either gets lost or becomes casuality his whereabouts can be traced even via satellite. In the case of battle casualties treatment given through out the evacuation chain can be endorsed in a compressed manner to be read by the 'hand held' computer at every successive destination. Often enough considerable time is wasted in organising the right type of blood etc., in an automated environment all such requirements and details can precede the casuality. Fed into the main computer it will help the doctors in giving further treatment with minimum loss of time, and thus save a large number of lives and limbs.

Factors to be Guarded Against

- With centralisation becoming a reality in a computer-based system the higher commanders wanting to start controlling the functional level logistics becomes a possibility. This could stifle the initiative of the immediate commander on the spot. For, he may never be certain of the higher commander taking away resources he was banking on for his own plans. This could happen either because of the departmental head or the operational commander exercising such an option without ascertaining the actual need of the man on the spot. We have to see that the computer does not become an instrument for breeding uncertainty at various echelons of logistics.
- Once computer become an essential tool for management of inventory it could happen that following the American penchant for acronyms we may also adopt their kind of thinking. Thus, the more common 'jeep' may become to be known as a four-wheel all-drive cross-country lightweight transportation system or FWADCCCLWTS. Nothing will put our clerks and storekeepers off the computers more than this variety of computerese.
- One subscribes to the theory that with computers we will attain cost effectiveness in administration, communications, material intelligence, and ultimately in operations. Many horror stories from Vietnam attest that the administrative overcentralisation (that is normally a consequence of efficient, fast paced accurate information and communications) led to gross waste and inefficiency. It created impossibly long

reaction times, and a loss of initiative at the lower levels of the military hierarchy. We should avoid saddling commanders with unnecessary and unwanted information so that they do not suffer from what could be termed as "information distress". The understanding of the nature and degree of control on logistics which operational command should exercise is essential to the attainment of combat effectiveness. It would be more pragmatic if the higher headquarters retain control, or remain aware of availability of a limited range of inventory only and selective reporting be adopted. This way the higher commanders will not be saddled with computer print outs they don't need and commanders at lower levels are not always worried about computers reflecting badly on their efficiency. In the system to be designed for 'inventory automation' it was recommended that out of over 5 lac odd items, the Army Headquarters should 'worry' itself with not more than five thousand items. It was felt that we should let the departmental functionaries and combat commanders deal with their specific areas of concern at lower levels. The computer need not become an "instrument of terror".

- An area of concern will be the fact of total dependence on computers, their operators and maintenance engineers. There will also a dependence on the "programme designers" who design the system. It could happen that the system malfunctions at a crucial juncture. This means that whatever be the value of the improvements, we will lose some form of ability to control the situation. With that loss will come some loss of responsibility. Because, when the "computer system" goes down responsibility shifts from the user to the repair technicians who will have to resurrect the machine. Similarly, if the problem relates to the software, the programmers must effect changes, because they are the specialists. This danger in over reliance on others is particularly acute for the field forces, which, often, operate in inhospitable areas and are widely dispersed, located away from the base support. The situation can become acute because the technical support from the 'hardware vendor' or the 'software engineer' may not be readily available in such places. Unless appropriate measures are taken in the initial stages the responsibility for fault free operation of the system gets abrogated. One way to partially overcome the problem is to have officers and men from a particular Service associated with the "system designers" from the very beginning. This team should subsequently train other users as well. This "in house" team of experts should be able to handle normal system 'breakdown problems'. They should also be able to identify and remove routine defects from the "hard ware". Another thing that can ensure efficient support in such cases is to keep integral loads 'light' and the facilities not immediately required should be centralised at locations wherefrom the stores can be made quickly available. This will help the support service commanders to tide over tricky situations when they arise. If the main system at the base can support operations till own programmes resume functioning that will be even better. In the "horizontal support"(discussed a little later) approach adverse effects of a system's breakdown will be considerably nullified.

Mere existence of a new technology is no guarantee that a panacea has been found for all our problems related to logistics support to the troops. Computer field today is highly dynamic. Every day some new development or product improvement is taking place. There is always a chance that one could fall into the 'technology trap' due to fast changing and improving hard ware and soft ware and thus be wasting a lot of money, time and effort without tangible gains to the organisation. Need of the day is that the Army must link the evolution of technology with its vision for the future. Technology and infrastructure available in the civil market will test the Army's ability to bring about and maintain a technological edge over potential adversaries. As has been put across in an earlier chapter we must integrate our defence technology needs with overall technological base in the country.

Notes and References

1. The term implies entire holdings wherever a particular item is held, its details are vailable to the controlling agency.

15

Asset Visibility

Asset visibility, a term roughly equivalent to inventory management, encompasses inventory in motion, including controls during transit of an item of equipment that are supposed to allow the sender and receiver to know the location and status of specific items in that inventory at any time. It also includes the management of material items at points of origin and destination.

It is an accepted fact that efficient and effective management of inventory throughout the supply chain significantly improves the ultimate service provided to the troops. This implies a high degree of assurance and responsiveness. Inventory control, asset placement, speedy restoration to serviceable state of the repairable stock, casualty evacuation to the right hospital and location of appropriate blood group etc thus; assume a significant status in providing a reliable and dependable material, maintenance and medical support to the troops. In so far as the equipment and weapon systems are concerned spare parts for the weapon systems, communication systems, machinery and other items (being in limited supply) there has always been a requirement for more accurate 'asset visibility' so that correct and timely provision action gets taken. In the existing system where reports and returns with regards to the material support are normally generated manually and dispatched by post the information is generally outdated. Signal communication with messages containing long winding part numbers and other sundry details that are required to be endorsed in relation to any material is often too garbled to help in decision making on the subject. Position gets to be more acute in a 'war like' situation when quick response is of paramount importance and the relevant item may or may not be available in the immediate vicinity of the area of action. To locate what is required, where and when required is crucial for providing an efficient and reliable logistics support. The data and correct details are important for this purpose. Our present system is not ideally suited for the reasons enumerated above. It does not provide suitable means for making an appropriate assessment of our current liabilities or future needs. Logically, how to provide an acceptable level of support has been causing concern for a long time.

In an ideal situation it should be possible to keep track of an important item through out the supply chain so that it could be made available to any user, based on urgency of his requirement- irrespective of his dependency. This means a total "asset vision" to the resource managers and controllers at all levels and their being aware of operational requirements at various locations so that correct and timely decision gets taken in making material available to the troops. Conceptually, what "night-vision" technology is to the battle-field "asset-vision" technology is to logistics. As night- vision technology has changed the way we train and fight, so is asset -vision technology expected to change the way we provide material support. Automation allows army to introduce asset-vision technology where the assets of all the logistics nodes can be linked into a "gate way" database. As a result, the depth and range of material available to the customer connected to that database can dramatically increase thereby giving him greater confidence in his support element, besides allowing him elbowroom and flexibility where his operational tasks are concerned. He will not be encumbered with too heavy an integral tail because his resource manager will be able to locate the necessary items anywhere in the vicinity and provide material support. With asset vision and inter connectivity, we can see more clearly where all items are located within the operational theater and outside its range of dependency. It also helps the 'supporting commander' in placing material resources in a manner where he does not saddle the combat troops with unwanted inventory, at the same time retaining the capability to meet both the predicted and unforeseen requirements.

Asset Visibility and Flexibility in Planning

Now with the availability of more accurate means of computer based simulations and intelligence it is possible to make more efficient logistic plans, and forecast distribution pattern much before any operation is undertaken. In a computer-based system the "resource manager" can evaluate and provide various options to the commander. He can also give him an objective opinion. In an environment of shortages, deficiencies in equipping our forces the "asset vision" be used as a medium for the commanders to develop and evaluate plans where the effective and innovative use of resources can influence decisions as much as the dictates of the operational situation. A higher commander can ascertain possibility of supporting a mission that had not been originally catered for, or exploiting a success. He can even simulate the data and use "forced decision matrix" to arrive at a reasonable conclusion with active assistance of his support troops. This means that the Service Heads at various organisational levels become active participants in the decision-making process. They must use their multidisciplinary skills to help create different opportunities that are based on sound operational, logistical and financial analyses. They must become advisors who can improve the operational effectiveness of their organisation by providing systematic, and objective scrutiny. All that the commanders at various levels need to appreciate is the fact that in the realm of material support to the troops an important input is 'asset visibility'. It allows a commander to make operational plans that are materially supportable and economically viable. (Louis C. Seeling in his book "Resource Management in Peace and War" advocates that in fulfilling the role of adviser, and as a member of the primary staff, the resource managers can be active participants in policy development. They should render opinions on policy development, rendering opinions on policy, procedures, and issues generated from all activities within the command even to the extent of evaluating operational plans.)

Responsibilities of Resource Manager

The linkages between the organisational structure and the resource management process, between the numerous types of financial and operational plans and between the approved resource allocations confront the resource managers with a number of responsibilities, particularly during operations. To meet these responsibilities they must ensure that an appropriate and effective control system is in place to execute the operational plans. The Army builds the logistics planning and technical manpower control processes around the development of operational war plans. Such plans are based upon the requirements of the tactical situation to ensure delivery of the necessary troops and equipment at the time and place required. A large number of inputs go into making a realistic estimate of the material support readily available and what can be made available within a prescribed time frame. This whole exercise will involve a number of different agencies, both integral to the military and outside its control. To get all the details and to arrive at accurate decisions quickly based on these inputs is at any time a difficult task. Particularly when today's environment is commodity oriented and features aging technology, procedures, and methods. Decision-makings are hampered due to incomplete, and often suspect 'item management information'. The decision making process is also hurt by limited 'asset visibility' because the means of communications for the Service units (generally) are rather primitive. The support Service commanders are often not aware of what is available and where outside their immediate span of control. Consequently, an unrealistic assessment and provisioning action gets taken which can lead to either too much or too little creating either a logistics snowball or a logistics draught. Both the situations are undesirable and unacceptable for successful operations.

Role of Resource Managers

Certain experts in the field of logistics feel that whilst there should be a balance between initiative and co-ordination in logistics operations, logistics commanders at every level should think ahead, while gaining empathy with the operational staff. It will be essential to take the 'Service Heads' in confidence if we want to reap full advantage of this powerful medium that makes "asset visibility" a more dynamic input than what it is today. In the absence of complete information regarding the overall plan he may place the systems and subsystems at locations which are convenient from storage and departmental functionality point of view but may not be located appropriately in so far as the operations are concerned. This means that the placement of logistics resources could be faulty if there is lack of communication between the supported and supporting commanders. This could result in resource and inventory imbalances that adversely affect the availability of weapon systems. This could create difficult situation, more so when the troops are deployed in different areas and in different roles, like defensive, offensive, holding, or "internal security" duties etc. Easy visibility of assets and an assured supply along the chain caters for the unforeseen because the commander in control of the theater will have information with regard to availability along the "horizontal material chain". He would have got the opportunity to seek approval of the ultimate controller of the operations with regard to the extent up to which he can marshal support from outside his jurisdiction.

Horizontal Logistics Support

An efficient and reliable 'strategic logistics system' requires having sufficient personnel, equipment and stocks on hand and also having the ability to re-supply and reinforce on a continuous basis. In our existing system of material support we go by the channel of base depots to the regional depots, to the divisional logistic units, and thereafter to the user units. The distribution is, thus, based on multiple echelons. The inventory is held in various ranges and depths depending on the "turn-around" time, relative importance of the operational requirements of the dependent units, essentiality of a particular item and terrain etc. This, results in an increased lead-time, enhanced transportation costs and manpower requirements. This is also instrumental in causing deterioration of stores while in storage or in transit and duplication of effort at various stages because we could have built up the stock in anticipation of its wastage pattern only. Actual consumption pattern does not necessarily match these predictions leading to either under or over provisioning. Thus making 'sustainability' unpredictable to a very large extent. Asset visibility will overcome this lacuna to a very large extent because whatever is held in excess in one theater can be quickly transferred elsewhere. At one time a particular "radio set battery" was a major problem in J&K because the sudden influx of troops had created a demand pattern that had not been planned for. Nor the suppliers were in a position to meet the requirement in a suitable time frame. During a discussion on the criticality of the situation the author could report having seen an abundant stock of the same item in one of our 'material support, units in the North- East (where from he had moved to Army Head Quarters). Presently this "holding" was serving no useful purpose at that point of time in that area. The higher commanders when told of this anomalous situation took immediate transfer action and thus saved the situation. There is any number of such contingencies where availability of correct and accurate information regarding stores could have improved the support, provided information was available to the logisticians.

The 'horizontal support system' can effectively reduce the size of inventory to be held by an "operationally committed" formation, particularly when short and intense conflict is envisaged. Because short conflicts require less maintenance; the equipment can be allowed to run down where replacements are available, alternatively assistance will be handy from sources other than those that are integral. Since, road and rail networks will also suffer less destruction and troops do not have to carry too much of rations and fuel we can manage with less logistics troops integral to the formation, unlike in a situation where a longer war is envisaged. Instead such support services troops (thus saved) be centralised within the theater under a higher commander who can make a greater use of such pooled resources. This will be particularly useful in the case of specialists and maintenance engineers where a higher commander can ensure a better equipment sustenance and maintenance. Integral logistics units will also not have to carry all kinds of spares too far ahead in the area of actual conflict. This can only be done if we have 'seamless' logistics support to our field formations. That means 'asset visibility'. In order to provide "asset visibility" and to make a comprehensive and appropriate use of the computers for providing logistics cover to field formations we will first have to identify specific areas where the power of the medium can be gainfully utelised. There is a need to make our distribution system more responsive to the user's requirement by ensuring that the right quantity and quality is delivered at the right time, right place and right cost. Efficiency and economy can only be ensured in an automated environment while meeting this requirement. This can be achieved by making troops dependent on 'horizontal logistics nodes' in addition to their integral logistic units. We could, also adopt "transportation model" wherein the stores are consigned directly to the regional depots or to major consumers directly by the suppliers, thus saving on the handling and preventing breakage en-route plus making the material available faster. There is hardly any logic in sending coal, petrol and diesel first to the 'supply depot' and it being transferred to a bulk user therefrom. For that matter why should a base ammunition depot be sending consignment of Artillery and Armour ammunition first to a regional ammunition depot and the user drawing it thereafter. These situations are not uncommon even where the actual user is located in between the two.

What is the "Transportation Model"?

Transportation Models basically define a distribution pattern where stores are dispatched directly from the manufacturer to the user. A base depot could also employ this method of sending consignments to the end user directly. Transportation models are becoming popular in more advanced Armies of the world. Trade makes regular use of the system for faster conveyance of stores to far off destinations. There are certain basic requirements for its adoption in our Military. Primarily, in a transportation model there are a group of origins (factories & base depots) and a group of destinations (users) and each origin has a capacity and destination. There are, however, a few conditions that have to be met for effective use of the transportation model. These are:

(a) Items originate from multiple sources and are to be transported to multiple destinations. A computer-based system where all the details regarding the suppliers and demand details are available can give relevant inputs for directing consignments. In fact, in the case of big consumers with limited storage space one can even pace out the delivery schedules so as not to cause any cluttering up at the user's end.

(b) Time/Cost benefit (both in terms of financial costs and opportunity costs) to transport one unit of the item from each source to each destination is quantifiable. This will help in deciding whether to adopt a transportation model or to go in for more conventional means. Will be particularly useful in the case of ammunition items, fuel, and bulk movement of the ration items.

(c) Each source has a specified quantity/assured capacity to supply the item. Level of assurance has to be high for a transportation model to be successfully implemented. Once a computer based connectivity is established between the supplier and the demanding agency it will be possible to regularly monitor the movement of stores all along the supply chain.

(d) Each destination requires a specified quantity of the item. It can happen that shifting of the troops or change in equipment profile leads to a situation where the demand for a particular item undergoes a significant change. In such a situation the monitoring agency can moderate the supply pattern well before the stores leave for their destination. Will be particularly useful in a real operational situation when formations can shift from one axis to another and commensurate material support is not likely to be available at the other end.

(e) The item is discernible in terms of standard units such as wagons/tons/packages or cubic meters.)

The above list is by no means exhaustive. Depending upon the usage, level of assurance, criticality, and economy of effort any change or modification can be made by the controlling agency. Almost all the big companies in the private sector employ transportation models while dispatching stores to various destinations. In a limited way Army also uses this pattern of supply. In an automated environment it will become easier and more economical to do so.

Efficient 'asset visibility' would also mean tremendous economy to the state. Because, in so far as the costly items are concerned the fact that an item is stocked in the horizontal chain would affect our criteria whether to add that item to the integral logistic unit or not, thus leading to economy of resources of a very high order. An automated logistics support system allows us to introduce the criteria that new items be only added where actually required. Common user items are to be added only on critical assessment of demand and supply pattern. Data simulations, suitably moderated, can make our future requirement predictions more accurate. The flow of information upwards, sideways and downwards will be precise, accurate and open to immediate feedback with computerised support for prioritising and analysing. Thus, 'asset visibility' at various echelons with inter connectivity will cater for a more pragmatic and realistic assessment of actual needs to be met immediately and those which could be met in a prescribed time frame from regional and base channels. This comes in particularly handy when some specific item(s) is in short supply in one sector but held in abundance elsewhere. The present situation; where it is not uncommon for one workshop in the theater wanting to back load some spares being surplus to their requirement but urgently required by another workshop in the same area; will also get straightened. All round efficiency and availability of equipment can be improved with the 'strategic logistics' plans becoming more realistic and reliable.

Logistics Responsiveness

For an operational commander the basic issue is that of responsiveness from his logistics elements. This requirement of responsiveness of a logistics system is intimately related to the study of usage data, planning factors, the "logistic snowball", to flexibility and to "readiness". In a computer based system it is possible to integrate all the factors for exercising control at an appropriate level. But sometimes, individual responsibilities- or processes are neither very clear nor well defined. This creates an environment of uncertainty because such ambiguity leads to either misplaced initiative on someone's part or unnecessary delays in meeting an urgent requirement. A large and complex organisation such as the

Army, therefore, needs a rigorous management control system to ensure that each staff element performs its tasks in a manner that provides the troops with all their essential requirements in an economical and more reliable manner. Logistics Assistance Networks can be set up in every operational command. These networks must be able to communicate accurately and quickly with forward troops as well as the base depots. They should be able to send regular updates on the material support position both to the customers and suppliers. With 'asset vision' becoming a possibility we could also set up "Combat Support Networks" at Corps level.

Combat Support Network (CSN)

A password protected web site can be created where the logisticians can provide continuous information with regard to supply position, technical and maintenance information to the users. A commander can more effectively monitor the flow of material in transit. A more vigorous control can be exercised at traditional bottlenecks like railway transshipment points, ship unloading and those transfer points wherefrom the stores are transferred from one carrying agency to another. It will also be possible to divert stores in transit to another location where an urgency suddenly occurs. It can use plug in programmes to clarify complex maintenance and logistics problems. It can incorporate videoconferences where the user can discuss equipment-related problems. In addition to providing real time data CSN will also reduce costs and efforts involved in typing letters, sending reports, using postage etc. It can quickly download EMERs (Electrical and Mechanical Engineers Regulations- these regulations generally define specific repair and maintenance charters and form an authority- along with certain others- for demanding spares, oils and greases etc.) to all the users of a particular type of equipment thus ensuring speedier procurement of spares, oils and lubricants required for maintenance. Similarly, in the case of any newly introduced equipment instructions can be quickly downloaded to all concerned. CSN should, in due course of time, become the hub of all logistics related activities within a theater and outside. It could, however, create temptation for centralisation of all controls. An important factor to guard against is that with network connectivity and better means of communication there could always be a tendency and a temptation to centralise such a control, ostensibly to make a more judicious use of resources. It is important to remember that the larger, or more centralised, our over-all system becomes the greater will be the need for responsiveness, and the greater the difficulty in attaining it. Therefore, there is always room for a great deal of thought and pragmatism to annalyse the situation and necessity of controls so that the economic advantage of centralisation does not convert itself into a handicap in due course of time where personalities instead of operational urgencies start dictating distribution of material.

Effective Use of Internet

If we quickly 'bar-code' our material and medical support and put this data on the Internet where password protection can ensure secrecy we can have all kinds of advantages. Commanders and logisticians can keep track of material in motion, in a critical situation a higher commander can be conveyed exact requirement. He in turn can locate the items anywhere in the country and select the site closest to the area of operation for ordering immediate transfer. In a case involving medical emergency relevant experts can be kept track of and their service requisitioned without any delay. Such experts can even help and advice doctors on the spot by 'video conferencing'. Similarly 'equipment experts' can assist in speedy diagnose of a failure or subsystem damage occurring at more than one location. A 'bulletin board' can keep every body informed of operational developments. Where decisions with regard to some equipment under trial are concerned the unit carrying out trials can interact with the developers and manufacturers and seek specific information in double quick time.

The chasm between the conventional logistics support system and a true result oriented technologically supported system is not going to be easily crossed. Organisationally we should want to be on the other side in order to make the leap successfully. For this to happen we have to be organisationally adaptive. But then military as an organisation has a lot of natural inertia, which inhibits any change in direction from what is described as a "time tested system". Computers being somewhat new for some of our commanders this inertia tends to increase with the ever-evolving computer technology making inroads into our material support system. No one should expect that it will be possible to plan perfectly in war even when we have computer-based systems in operation assuring 'horizontal support' and providing a total 'asset visibility'. Nor will it ever be possible to eliminate selfish, deliberate over estimates because a computer cannot on its own carry out any supervision or check on these human tendencies. Precautionary hoarding, in any case, is a national trait; and there is always a natural desire on the part of each man and leader in combat to have an ample factor of safety in his resources. It is only when we, the logisticians, can give a commander confidence in the quality and adequacy of his planned logistics support that these habits of excessive over planning, over demanding and over provisioning will break. It is for the logisticians to convince the unit commander that his real inventory lies within the system itself or he will encourage the establishment of larger than optimal inventories to maximise his own operational readiness as he sees fit. It is felt that if these practices could be controlled and reduced, and if inadvertent over planning could be reduced, tremendous savings can accrue. One factor that could cause some worry is the fact that 'formation commanders' are usually protective towards material resources that are authorised and held by the logistics units under their direct command. A demand meeting system, where utilisation of 'horizontal inventory' is envisaged, could cause heartburn or even a sense of insecurity on the premise that the item may be needed by own troops. With more accurate predictions and procurements and consequent improvement in supplies such reservations will soon disappear. It will be good for the 'combat commanders' to remember that a military force designed logistically for a short war can usually also fight a long war if necessary, provided long lead time items are hedged and economic resources are available. A military force which is structured for a long war and which is consequently weak in short war capabilities may not survive to marshal its potentially superior power if it is loaded with too much of material support. The criterion will not be the ability to sustain every unit all the time, but to support the total force as required. A computer-based system will ensure a healthy blend and balance between the two requirements.

16

Integrated Logistics Support System

Most countries amalgamate their military needs in their industrial, technological, financial, trade, and other developmental planning factors while taking major policy decisions that have a bearing on the economy and industrial growth. This way they can ensure a healthy blend of population's requirements with that of the military. Thus, making optimum use of the available resources. This also ensures that the future plans in all the important fields take the military needs into account before starting any groundwork or making policy decisions. The problem is that unlike .USA and certain other advanced countries we do not have a national level body dealing with 'logistics planning' that could keep defence requirements in perspective while allocating national resources. Within the three Services also there is only a limited commonality of interests in so far as achieving material sufficiency is concerned. This leads to duplications, delays, missed opportunities and wasteful efforts. For example, Army, Navy, and Airforce have undertaken "inventory automation" separately inspite of commonality of procedures and interest in providing better service to the troops. They could easily have gone in for a common system and saved money, besides ensuring a more effective response to the user's requirement. Paradoxically, logistical decisions are not always guided by any sense of organisational economy; or concern for the poor 'tax payer'.

Consequently, dissatisfaction prevails because the "wasteful" expenditure incurred in safeguarding' the "empire" leaves that much less money for meeting material needs of the troops. Some kind of steps in the direction of achieving economy and efficiency in logistics are taken from time to time in a manner that could at best be described as half-hearted. More often than not specific Service interests tend to dominate decision-making. If not that, then the 'bureaucratic finger in the pie' ensures that the three Services continue to play their separate tunes. In so far as the Army, per se, is concerned it is not uncommon for a Support Service to be wanting to either go in for "empire building" or wanting to be one up on the others, often at the expense of organisational interests.

The present practice of overpowering the logistics mission by deployment of new technology alone (as we have been doing hither-to-fore) is no longer viable. The real challenge for the coming decades is to develop new ways of satisfying material requirements as against attempting to modify old ways to work more efficiently. Many traditions and practices that characterise today's logistics structure are both archaic and symbolic of a bygone era. We, therefore, need to examine basic concepts that are vital to the emerging form of logistics operations. There are a variety of approaches available which can help meet future requirements within the philosophy of 'integrated logistics' so as to make the system more responsive to operational needs besides bringing in economy. For one, coordination of inter-organisational logistics performance provides an opportunity to eliminate duplication and minimise risk for the total channel. This means that doing more of the same things in logistics, which we have been doing in the post independence era, is not the answer. At the time of independence our Airforce was too young and the Navy was too far away from the conventional logistics system to be given an efficient material support. Over the years with an improved rail, road, air and communication system a time has come for us to give a total re-look to our logistics operations. An integrated approach will help provide economy of effort and better resources for meeting operational needs. Two models which represent a new or different way of formulating and conducting logistics operations using currently available technology with a potential to absorb future technological advancements merit consideration.

Integrated Logistics Support Structure for the Army

The term "Integrated Logistics Support" in itself has two connotations. One, the entire aspect of logistics complement for particular equipment or a weapon system is examined in its entirety as a package. Two, various Services are integrated under one head. The concept is of a rather recent origin. Traditionally, defence forces acquire a weapon system or major equipment/plant and then begin to think of storage space, material handling equipment, and maintenance of these weapon systems/equipment or in other words, the logistics support necessary for the same. Thus, logisticians are brought in rather late in life to determine what maintenance, spares, documentation, test equipment, etc., are needed to support the equipment. What adds to the problem is the compartmentalisation of the support in a manner that defies logic. Resultantly, user is never sure of whom to approach for what requirement. Often, this approach reduces the effectiveness of the major equipment in some of the phases of its 'life cycle' apart from, sometimes, proving that the chosen weapon system/equipment to be not the most cost-effective in the long run, over the entire life-cycle of its use. This happens when the cost involved in creating some special storage conditions, or highly technical training, or some modification to the 'bought equipment' comes up. Defence Services, today, are becoming increasingly aware of the requirement for logistics support planning for equipment, which in itself is becoming technologically more and more advanced every day. Every effort is required to be made to ensure that weapons and equipment are introduced into service complete with various elements of logistics support .

All this really means is a requirement to integrate various functions in a manner that ensures effective management of the newly inducted equipment, bringing us to the concept of "integrated logistics support". The developing concepts of taking a "systems view/ systems approach" in such activities, resulted in the development of the concept of "Integrated Logistics Support" for the Equipment. In this the logistics aspects are considered right from the word 'go', thereby making logistics as "before-the-fact discipline" instead of it's being an "after -the- fact- discipline". This is done in three ways:

- Providing design guidelines concerning logistics factors e.g. habitability, safety, availability requirements, storage facilities, manning, and standardisation.
- Timely evaluation of the interactions and input of logistics elements to other considerations.
- Integration of logistics factors into total system design through trade offs.

The integrated logistics support is also capable of making full use of numerous available mathematical and statistical methods as well as "operational research techniques". Simulations, accelerated wear tear etc., are fed into computers to arrive at various decisions connected with the entire life cycle and deployment options are studied before firming up the operational and functional 'qualifications' requirements. This is an important facet of the term "integrated logistics support" where all aspects that ensure efficient performance of a weapon or equipment system are examined collectively. The only thing is that if the decisions are taken under one head before a weapon, communication, aircraft or even a Naval ship is physically inducted in service we could reap full benefits from the money spent by the country in equipping its forces.

Army Logistics Corps- Broad Discussion

Our present system, though conducive to a reasonably efficient support suffers from the uncertainties inherent and in built. Due to duplications, division of responsibility, frequent changes in the responsibility the user often suffers as these departmental changes and alterations do not reach him well in time. Too many agencies with their parochial considerations, and 'empire building syndrome' continue to influence decision-making their efforts may not necessarily be economical or in the state's interest. Besides causing delays, uncertainties and numerous other problems to the troops the present system does not lend itself to an environment where we could make optimum use of the latest technology in so far as the support to our forces is concerned. The situation also does not lend itself to "down sizing" or "right sizing" because a total and impartial view cannot be taken by a single individual. An ever expanding need for an economical material support makes it imperative that we go in for a change in our thinking if we want to ensure enhanced efficiency at a reduced cost and with reduced manpower. This is becoming especially important because of the increasing variety and complexity of the weapon systems, demanding greater involvement of the technicians, monitoring and follow up actions from the logistics support system. Our present methods of providing logistics support do not seem to be capable of meeting this challenge because these are not geared for technology changes and consequent challenges. Therefore, one feels that the concept of 'integrated support' has a better future if the "Services" themselves are integrated under one head.

Integrating Logistics Support Services

Certain advanced countries have gone in for an "integrated logistics support system". The term has two connotations, viz., a system that envisages a combined support to the Army, Air force and Navy. i.e. establishment of a 'Defence Logistics Service'. In the second case there is a system being adopted in some countries where various Services like Ordnance, Engineering (minus Combat Engineers), Medical, Transport and Supply etc., are brought under one head. That is, something like establishing a 'Logistics Corps' under a "Chief of Logistics" at the Army headquarters who becomes the adviser to the Chief of Army Staff on all matters connected with logistics, and pooling Services under one head at Command level and lower formations down to a Division. Broadly, we will discuss both ie Defence Logistics Service and Army Logistics Corps. A change in our thinking with regard to logistics support to our troops has become imperative. The present method of material support where we continue with the system developed by the British and when the communications were rather primitive is outdated. Another worrying factor is that we have gone ahead and created so many directorates and mini directorates that as a user one is often not sure whom to approach for what purpose. Some of the Services (like Army Education Corps) have become almost redundant in today's military. For that matter "Value Engineering Directorate" does not seem to have provided any particular value for the money expended in the effort of creating this directorate. A brief discussion on our present system is called for before entering into a resume' on the Army Logistics Corps.

Our Present System

In the Army, logistics support functions for the "in-service" equipment are carried out by the AOC, the ASC, EME, and infrastructure and buildings etc., by the MES under the aegis of Engineer- in- Chief. Medical service, Postal, and certain other logistical functions are performed by the Adjutant General. Thus we see that the agencies responsible for providing logistics cover are organised under four different PSOs at Army Head Quarters. This means that the management and control of the Logistics Services are not under a unified single management or control. The staff support is also highly fragmented and does not approximate to the concept of an integration of the material service provided to the combat soldier. This, often, leads to a situation where a new 'weapon' or 'communication' system arrives and there is no provision for its proper storage, neither at the user's end nor, for that matter in the depots. Even ammunition, often, lies in the open under 'tarpaulins' for years, losing on its "shelf life". Costly missiles are kept under "make shift" arrangements while various agencies take their time to decide where and when to make appropriate accommodation, or who should undertake repairs, testing and periodic maintenance etc. This is so because in addition to various other factors; in the case of "works" after moving through numerous channels and screening at the "Engineer in Chief's" branch the proposal is again subjected to a detailed scrutiny in the "Quarter Master General's" branch by "Q works". Far removed from the scene of the problem, the decisions taken with regard to providing appropriate "covered accommodation" to this expensive material procured, often from abroad, lacks care and consideration. Such duplications, and often triplication, leads to extended "administrative lead-time" consuming years and years in the process. This leads to uncertainty, inefficiency, and consequent cost escalations, inadequate training of the technical support personnel, and (some times) premature down-gradations of these stores. This kind of unimaginative approach to the serious subject of economy in resource utilisation causes an apparent loss of confidence of the user in his 'support backing', for no fault of the Support Services. Another major area of concern is that even support services tend to look for opportunities for expansion when they know that the existing infrastructure could absorb new equipment. Typical case is that of "missile testing". There have been any number of changes in decision on the subject. Depending on the vocabulary, marketing skills, old boy nets the responsibility has, alternately, been given to 'Ordnance or EME.

In the case of "Air Force", the "Logistics Branch" performs all the materials management and physical distribution functions. The maintenance of the equipment is carried out by AE (M) and AE (L) branches. These agencies are organised under the Air Officer-in-Charge (AOM) at Air HQ, thereby providing a single point management and control of these activities. In Command Headquarters, the logistics and maintenance branches are placed under Senior Maintenance Staff Officer (SMSO). The Command Logistics and Maintenance Officer (CLMO) is responsible to SMSO for all the logistics activities of the wings/units within the command. The organisation at wing/unit levels i.e. at Air Force Stations does not follow the unified command and control concept as it is at Air HQ and Command HQs. Logistics support for the newly introduced aircraft systems and equipment, is provided by the constitution of "Initial Provisioning Committee"(IPC)/ "Maintenance Planning Teams"(MPT). This arrangement is, in fact, more or less an "after-the-fact" approach. Integrated Logistics Support, in the true sense, does not as yet exist in the IAF even though some attempt has been made to integrate logistics support for the Jaguar Project.

In the Navy, the 'Executive Branch' and the 'Technical Branches' are now performing materials management and physical distribution activities, which were earlier performed by the erstwhile Supply and Secretariat Branch. These agencies are organised under the "Chief of Materials" at Naval Headquarters thereby providing a single point management and control of these activities. Logistics support for newly introduced equipment, however, is planned and organised after selection and ordering of the new equipment by the user directorates and hence is again an "after-the -fact discipline". Thus, integrated logistics support, in its classical sense does not yet exist in the Navy either.

As seen from a very broad 'over-view' of our present system given above the need for a speedy over-haul and change over to a more responsive and economical method of ensuring a higher level of support and response is evident. In order to ensure a smooth transition it is recommended that in the first instance the Army should amalgamate these functions under one head. Thereafter, and over a period of time Services that are common to Army, Navy and Airforce be brought on a single grid.

Case for 'Army Logistics Corps'

Presently the logistics support for both war and peace is influenced by what we primarily inherited from the (then) existing British system (British Army has since gone in for the Logistics Corps). The inadequacies of the system have broadly been discussed earlier. To overcome the mismatch between the demands of time and an efficient logistics support there is a case for an "Army Logistics Corps". In this all the services (including MES, Medical, all kinds of military procurement etc.) are brought under one head responsible to the Army Chief for their efficient functioning. This "Chief of Logistics" will work in direct concert with "Director General of Military Operations" and plan the entire gamut of material support. Because the entire exercise will be controlled by one commander instead of four to five PSO's now dealing with the subject of logistics there will be speed in action and instead of a compartmentalised and therefore perhaps a biased view the higher commander will have a total and dispassionate analysis of a given situation. Coordination between Army and other two sister services will also be more pronounced because it will be possible to quickly conclude whether military resources meant for a particular situation could absorb need of all the three services or not. A computer based MIS and Coordination Section will provide the "Chief of Logistics" with necessary inputs for decision making. He will have all the "Heads of Services" including 'Medical' directly under him. Procurement of all the stores will be centralised under a 'Director General of Procurement. This Director General will have under him officers and staff from all the 'Directorates' presently doing procurement of stores for their respective Service. This is to ensure that the 'domain knowledge' is available for advice and purchase of correct items at correct time and place. The ' projected figures' in respect of quantities to be purchased will be vetted for approval and concurrence of the 'Chief of Logistics'. Special accommodation for equipment and subsystems can best be coordinated if MES or the complement of Engineers involved in construction work is also brought under the Deputy Chief of Logistics. Additional Director General (Movements) will also function directly under the Deputy Chief of Logistics. This will ensure that availability of transport, railway wagons etc. are distributed as per the priorities influenced by " operational requirements" and as directed by the organisational head. Since, accommodation and movement of stores will affect all the services these two organisations will provide better service if their functioning is coordinated alongside projects and equipment introduction. To save on 'bureaucratic' lead-time it is recommended that a "financial adviser" be also made an integral part of this set up.

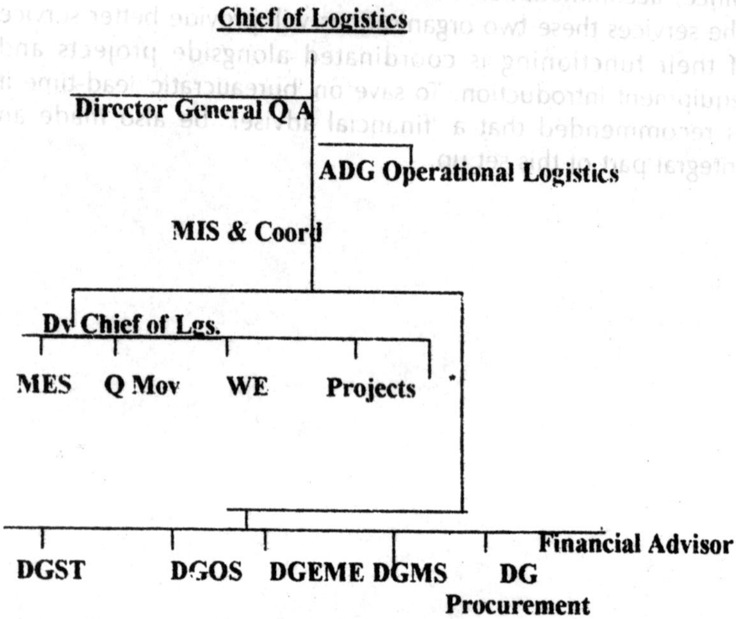

*All the minor groups like 'Equipment Management', 'Standardisation', Value Engineering etc be grouped together under the Deputy Chief of Logistics.

Pioneers can be merged with the Corps of Engineers; Military Farms, and Veterinary Corps with ASC; Army Postal Service can form part of AOC. And directorates like Quartering directorate can be dispensed with. (This Chart, however, is only a broad thought process). Presently, we have created any number of 'project management' teams functioning under different controlling agencies. This leads to problems of coordination, day to day monitoring, uncertainty, and bickerings. By putting WE directorate and project management under the Deputy Chief of Logistics we can ensure speedy interaction between the agencies involved in carrying out trials and introduction of equipment into service. By bringing all the activities (other than physical procurement, technical maintenance, and medical etc) involved in making resources available to the troops under one head we will be considerably reducing expenditure in carrying out these tasks. It will be seen that the organisation as suggested does not have any place for the 'Quarter Master General' and the 'Master General of Ordnance'. The new organisation can, therefore, be easily created without any increase in the number of "General Officers".

Following advantages are envisaged:

- Single manager concept becomes operative. "Operational Logistics Directorate" can be the overall coordinating agency. It will work in concert with the "Director General of Military Operations', thus saving on time presently expended in obtaining concurrence of various agencies involved in providing material support to the troops in forward and operational areas.
- Presently, the budget is allocated to different 'Directorates' for procuring stores. Often, they are not able to spend their full allocations due to various reasons. By centralising procurement it will be possible to make transfer of funds well in time for making constructive purchases and avoiding any 'last minute' money spending exercise.

- At present any major construction project after going through various levels of checks and approvals, including by the QMG, is again put under scrutiny in the "Engineers in Chief's" branch. This is a wasteful duplication. Under a single head such like duplications can be eliminated. Interaction between different 'Services' becomes more intimate and therefore more responsive to the user's needs.
- By routing all foreign purchases through one agency it will be possible to ensure that the requirements are consolidated and negotiated with the suppliers in a manner that gives the best price and time frame for the stores to materialise. Availability of the "financial advisor" will ensure that "administrative lead time" is considerably curtailed. Thus, avoidable cost escalations which are a bane today will be automatically reduced making army purchases more cost effective.
- It will be possible to build in indigenisation and vendor performance cells along with standardisation. This can bring in more competitive bidding and reduce reliance on foreign suppliers. Making the 'Chief of Logistics' as Chairman of "Ordnance Factories Board" may bring in greater accountability in so far as procurements made from the PSUs is concerned.
- The system could be a forerunner for an integrated system for the three Services. Thus it will be more dynamic and economical in the context of overall national requirements.

There are certain disadvantages also which could cause some problems in the beginning. Span of control for the "Chief of Logistics" could be considered to be too large. This means that 'automated MIS and energetic coordination groups are essential for smooth functioning. Although, the DG procurement will have advisers from the concerned Services, the thinking of some of those who may prefer to further the interest of their individual Service rather than that of the organisation may sway him. It may be a better idea to rotate the post of 'Chief of Logistics' among the three services viz. AOC, ASC and EME. Centralisation of procurement under one head may dilute the accountability of the Service Heads who deal directly with the suppliers today. This can be dealt with while fine-tuning the integration. The Service Heads if they project their requirements realistically and in time can be tasked to forward periodic reports on the stores availability to the Chief from time to time.

Such a system where 'Logistics Corps" has been created is operative in certain other armies including UK. While examining the feasibility in our context we could certainly gain from experiences of these armies. Of course, our own culture, functional environment and operational requirements will be the ultimate guiding factor. Enormous amount of money can be saved with enhancement in efficiency. A detailed examination of the proposal will reveal that the model has an in-built capacity for reduction in the number of "service troops" without any excessive strain on the efficiency of the services. Better coordination and speedier response to the user's requirements will give greater confidence to the supported troops, who sometimes (and rightly so) feel that there is some inadequacy in the service provided to them and the tax-payer is not getting his money's worth.

Defence Logistics Service

In today's environment where major operations, in all likelihood, will involve the Army, the Navy, and the Air force it might be more pragmatic and economical if an 'integrated logistic support system' is brought into use. All the three Services have their separate infrastructure to support its troops whereas there is tremendous commonality in the service provided. In remote areas and due to functional reasons common facilities for purely need-based reasons, in any case, exist. But these are exceptions where no body expects any shake up of the empire. Having a "Defence

Logistics Service" meeting all the requirements other than those that are peculiar to a particular service has not as yet been considered as a serious proposition at any decision making level. Having such an organisation will be a good thing for our military. Its main advantages will be:

- Flexibility to commanders in shifting specific resources to a more pressing need. The thrust would be towards ensuring that in order to provide responsive and flexible support the resources are placed in such a manner that any Service can make use of the same. It will need setting up of a 'joint machinery' at the highest level for allocations to be made for specific missions. To cater for the unforeseen situations cushions could be kept under central control. In this the logisticians must anticipate needs and not to wait to react to demands. This means that for a responsive and flexible logistics support the logistician's knowledge and understanding of the Commander's mission, his concept of the operation and his expectation of how and when he will fight the battle and the level of involvement of the other Services will play more important role.
- An integrated programme will provide better asset visibility for the logistics Services to plan, programme, and fund and operate the military logistics systems more efficiently. Setting up of maintenance workshops, repair and spares backing for common equipments, and ammunition etc., can lead to considerable economy of effort. It would, thus, go a long way towards saving money and resources. It would allow a more effectively supported force to try out new ways of logistics cover in all kinds of operational commitment and deployment.
- Integrated planning and control of the movement of supplies results in more responsive re-supply and smaller local inventories. Having fewer supplies on hand means reduced care and preservation efforts, smaller storage areas to protect, and easier inventory

management, meaning lesser number of men and a smaller tail at different levels. The integrated system will allow both organisational and direct support maintenance, to become the responsibility of a single higher commander. The Army, Navy, and Air force, though existing separately, would no longer provide their own support. They would lose their logistics "tails", which would be integrated into a "defence logistics service". This means that more combat units can be supported in combat simultaneously with lesser number of logistics units.

- High ratio of combat strength to support can be obtained by centralising the logistics support; and integrating as much training into the logistics structure as possible, keeping peculiar requirements of each Service in mind. Losses (casualties) among support troops will also be much smaller because of their being lesser in number present in the area of operation.

Units and formations with extensive organic support tend towards an expeditionary force posture that relies on its own resources rather than upon any local assets that may be available. Decentralisation and distribution of logistics assets to commanders preoccupied with an urgent combat task inevitably leads to a slighting of logistics and an attitude hostile to the uncertainties inherent in mobilising local resources. If, however, logistics are centralised under specialised logistic commanders whose attention is focused primarily on logistics, openings can be created whereby commercial or other assets can be used to advantage. Particularly in the likely areas of our operations there are many resources within the civilian economy that can be tapped in wartime to supplement and flesh out active forces. Transport, medical, engineer construction and signal communications are some examples of such useful logistics capability boosting civilian assets.

A defence logistics Service that acquires and provides equipment and sustains the armed services would also go a long way towards achieving the goal of fighting together. It will be important to ensure that differing nature of naval, land and airforce requires certain parts of their logistics support that conforms to their special characteristics. These elements should be tailored to fit these peculiar requirements of each Service.

If we create a combined logistics Service we will be able to establish a 'Strategic Logistics Agency' on the lines of US military. This agency could be set up under the aegis of 'Defence Planning Staff' that should work in concert with the "Planning Commission', and be tasked to undertake the following:

- Forecast the environment of 2010 and beyond based on an analysis of long-range planning guidance and future war-fighting doctrine. Projection of material support parameters for the future should stem from this forecast. This will help in making allocations for research, design & development, and creation of facilities well in time. Costly projects can be suitably paced thus avoiding a sudden strain on the economy.
- Determine future logistics opportunities by examining completed and on going studies now being carried out by numerous agencies and training institutes. A special cell can be created for the purpose where scientists from the DRDO, civil, interact with military professionals and work on desired future models. They could conduct exhaustive interviews with military, civilian, and industrial leaders with a view to identify potential applications of technology in enhancing operational capability of the Military. Industry will also respond more enthusiastically once it is known that instead of far too many agencies with varied concepts and ideas with regard to military needs they will be dealing with only one. Identification of a potential "bulk buyer" giving his time frame spread over a number of years will allow the industry also to pace its schedules in a manner that helps industrial growth along with an enhanced military capability.

- Decide on logistics foundations for the year 2010 and beyond by linking information about the future environment and logistics opportunities with information on logistics imperatives, air and land battle tenets, battle-field sustainment functions, and functions of the industrial and sustaining base. Where required, restructuring and redefining the tasks and roles can be concurrently undertaken.
- Develop plans to link validated logistics concepts with a long range integrated (Army, Navy, Air force) logistics plan. This will help in generating an environment where the Military needs get built into a "National Logistics Grid". This would mean that the defence budget, which today is taken as a "load" on the national economy, would become a part of national 'industrial and technological growth' and development. Thereby, generating jobs, business, revenues, technological development, and over all growth of the national economy. Improvements in our rail, road, and air communications can be examined in the light of their military utility. This will in publishing a 'broad-based' concept for logistics in the year 2010 and beyond. This should in turn help in identifying the technological mismatch insofar as the available, desirable, and possible capabilities are concerned. This exercise will help the decision-makers in deploying national resources in a direction that allows India to achieve self-sufficiency in all those field that help bringing in growth along with an economically viable defence capability. Making defence as a part of a national logistics grid will help in integrating military requirements into the distribution cycles and transportation models of our bulk suppliers of a very wide range of spares and other common usage items. This in turn will help in down sizing our service units besides bringing in a higher level of assurance to the users. For example, a manufacturer who has been

tasked to supply any specific item can straight away send it a designated regional depot or even to a support unit located in forward area bypassing interim agencies that can now work with a reduced workforce.

Establishing a Defence Logistics Service offers several advantages. One advantage will be in our whole approach to the subject of defence reorganisation and restructuring, planners will be able to take a total view of civil and military needs and make appropriate allocations. Since logistics is the least glamorous field there is likely to be very little resistance from operational angle. A common logistics system cannot be put together from our existing facilities. As has been discussed earlier, there are too many 'gray areas' and mismatch in the parameters exists. A common system would need to be created through infrastructure funds, and for military reasons rather than because there are considerable economies to be grasped. That is, any major cooperative initiatives in this area must be justified on military grounds rather than on economic grounds. Secondly, if economies and close integration are desired, it is necessary to take steps to achieve them, e.g. the use of 'management, costing, and operational research techniques to identify those uneconomical activities which impinge on the theme of a vibrant, economically viable, industrially supportable defence. As a first step we must identify those uneconomical activities (like economically inefficient and low technology PSUs) and do away with them. These are doing no good, either to the economy or to the soldier. In today's world where bottom line is the main strain and resource mobilisation the main theme, it is essential that we bring in a system that allows us to make plans that mean economy both in terms of money and effort involved in equipping our troops. Be it for war, or for peacetime standing military.

17

National Logistics Grid

Wherever in the world defence needs have been integrated with the country's 'civil end' requirements, the over all progress has been fast and phenomenal because it allowed participation of willing investors. In these countries identifying a specific military need and also finding a civil end usage for a 'system' or 'sub-system' has helped both the defence and civil industry in a big way, particularly where products in the fields of electronics, avionics, automation and transportation are concerned. It has encouraged adoption of newer technologies much faster and has helped the industrial growth. A poor country like India with too many problems can certainly make use of these observations and bring our military logistics on a common platform along with the common man's requirements. This will help us in providing material support to the military more efficiently and effectively and in an economical manner. Of late, some kind of effort in this direction is being made. The results haven't as yet encouraged the military or the industry. It is time we accepted the fact that doing so (that is, bringing military and civil needs on a common platform) in an unplanned and haphazard manner and by fits and starts is not going to meet the ultimate concern of providing a viable defence with less expense. Logistics cost a lot of money, they involve industry, manpower, raw material and virtually every walk of life is affected. By making military needs as the basis for our technological and industrial advancement we can have competent defence along with development. This can be done if we create a "national logistics grid". Before discussing the framework for such a grid it will be appropriate to examine various factors that go into creating an environment of efficient, industrially modern and economical national system that goes into providing logistics for the military.

Fundamental Elements of Logistics System

There are, at least, three ways to promote economic efficiency in the military use of resources; better understanding of the nature of the problem, timely and systematic quantitative analysis of military choice; and improved institutional arrangements Logisticians will always face pressure to accomplish more with less. They are required to plan in advance to optimise budgets and resources in such a manner that troops get what they need at all the times. The demands of economy both from the standpoint of the national budget and from the military principle of economy of force dictate that logistics support be furnished with the least possible waste and in such a way as to develop the maximum fighting power. All that it implies is that the objective of all logistics effort should be the creation of continued and effective support of the combat forces. We also have to remember that, whereas economy is essential to the attainment of this objective, economy in itself, is not the objective. All measures affecting the control, development and/ or induction of new equipment, and the coordination of defence logistics must be judged by their impact on sustained combat effectiveness under war conditions rather than by the sole criterion of peacetime economy. There are prescribed fundamental elements that go into providing an efficient and economical logistics system. These are people their level of technological currency, raw material & industrial base, relative locations, climate, weather and geographic factors. These elements create a "basic national economy" which consists of a "Political System" and a "Social system" which, animated by intangible motivation and desires as related to moral values, combine to provide education, science & technology, law, financial and industrial management, money and credit. Also agriculture, manufacture of material goods, investment, basic processing, transportation and distribution that in turn create a "Gross National Product". This gross national product is applied to:

- **Defence Needs**: Men, material, facilities and services less the portion applied to the para military forces and that fed back into the 'basic' national economy via procurement. These defence needs are analysed at an appropriate level and depending upon financial viability and other considerations like 'time and space', determination of requirement, procurement and distribution aspects etc., are processed accordingly. Every military need can be approached in terms of facilities, services, materials, and men. It will involve planning, organisation, execution and supervision.
- **Civilian Needs**: The civil population has to exist in a reasonable manner and they will require goods and services, transport resources, medical services and various other facets of every day life. Whereas during or prior to an impending operation certain deprivations can be expected, one can not have the citizens denying themselves for an unspecified duration just because a military need exists. In a democracy only surplus, nonessential civilian consumption can be diverted to military efforts. Other than that, defence needs will mean a strain on the national resources unless suitably provisioned for. In India there is no consensus among the politicians, bureaucrats and the economists on the mechanics of inter play between defence industrial growth and national economic development. At best the macro economic consequences of the 'domestic defence industry' have been addressed in a patchy manner. Consequently, we are presently in a 'no win' situation where failure in development planning is attributed to the increase in defence spending.

A country like India can absorb most of the "ailments" germinating from the present situation. Problems generally arise when identified defence needs are reactive and are in relation to what our hostile neighbours have or are likely to have and how much their acquisitions can affect us. Then money has to be found in a hurry and possible source of supply has to be located in order to bring our forces at par if not on a superior platform. It could happen that while we are looking for a potential foreign supplier our own civil industry could actually be having capability to meet that specific military need. In the absence of a common defence and civil industrial base such situations do create difficulties at times and need to be streamlined in a proper manner at the highest level of policy making. Since, most military needs are high technology-oriented, there can be more opportunities to apply civilian technologies to arms and communication systems development and production. Where commonalties exist as is the case with clothing and general stores the task can be achieved much faster. An effort in this direction will bring down the boundary between military and civilian technologies. With our healthy industrial and technological base there should be no reason not to blend our military and civil needs and achieve considerable economy and technological advancement as an aim plus.

Case for a National Grid

Considerable savings in our logistics elements can be made by using facilities that exist and those, which can be developed within the common ambit of military and civil use. For ease of understanding we can take the case of batteries, tyres and tubes. The military holds huge reserves of these items at various levels of holdings after procuring these from the trade. These are required to be held as ' war wastage reserves' and to meet seen and unforeseen requirements of the military. In strictly economic parlance it means "locked up" capital which could be put to some more constructive use. This also means that a large fleet of civil vehicles which might be off-road for want of batteries, tyres and tubes (now held as war wastage reserves by the military) could be put back on road and thus earn revenue for the owners, besides creating more work at different levels and funds in the form of taxes that are to be paid when a commodity changes hands. Such a move will also be ensuring speedy movement of essential goods. There is no reason, as to, why we can't lay down stipulations for the trade to supply certain quantities of identified spares and accessories when given specific and advance notice to do so. This way military can cut down on its scales of reserves. A suitable time frame for every type of industry that produces items used by both military and civil could be worked out. The trade should be bound to meet the defence requirements either by enhancing its capacity or cutting down on exports if they are already running to full capacity whenever "industrial mobilisation" becomes unavoidable due to military reasons. Thus, defence needs can be met without any undue strain on the population. This will also mean that the military will not be required to hold reserves of such items, locking up its capital which is always in short supply. Among other things it implies that for the greatest effectiveness and efficiency the military should make the maximum practicable use of civilian equipment and techniques.

National Grid in Other Countries

Consider the actions some countries have recently taken to increase the "linkage" between the military and civilian sectors. The government of the United Kingdom has used private venture capital to establish a firm called The Defence Technology Enterprises, Ltd., for the explicit purpose of transferring military technology to the civilian sector. Some of its 45 employees are located inside the Ministry of Defence workshops and laboratories, and their job is to identify ideas that can be declassified and used in the private sector. (Australia has set up a similar organisation.) France has formed a council, at the ministerial level, to address the subject of dual-purpose advanced research. Additionally, in 1987 France established a "working group to stimulate defence-civilian innovation" and set up a "special fund" for such activities. Italy has established an Inter-ministerial Coordinating Body for Defence and Civilian R&D. Japan is explicitly trying to integrate its growing defence production capability into its national industrial base.[1] All the advanced countries have not balked at bringing in their experts from whatever field to assist in defence effort without worrying too much about security, our common fear where it concerns any thing even remotely connected with defence in any manner. This appears to be so on the premise that whereas force deployment plans and other related subjects are 'defence secrets', the planning of production and supplies is not that sensitive an information.

In case we set up such an organisation where total requirements are examined and the industry is encouraged to participate, one clear cut advantage will be that the sizeable reserve of intricate spares and sub- systems that we have to keep today will not be required to be kept. The magnitude of savings in terms of hard cash will itself convince anybody that we must embark on such a course of action without any further delay. The military consumes an enormous variety of goods that are often similar to the same that are used in the civil sector. The needs are not confined to arms and ammunitions only. Activities in machine shops, foundries and laboratories, which exist in a fairly large number in both the military and the civil sectors sustain modern armies. A fairly large number of vehicles and communication networks are common. Repair and maintenance facilities, both for the military and civil customers, could be dovetailed by the concerned industrial houses in such a manner that the military could start making use of these arrangements to maximum extent and roll back its own infrastructure in major towns where the manufacturers have facilities for repair and overhaul. Military should retain only those repair complements that it might need in totally inhospitable areas and which are considered to be absolutely essential for operational reasons. This, besides generating more work and jobs for the civilian population, will help in reducing the tail element of our military. Similarly, development of the communication network in our backward areas could be (and it should be) combined so as to include both the military and civil usage. It will, thus, be possible to lead to over all improvement in these areas and help in bringing them into the mainstream and ensuring speed in movement of troops when required. There are many more steps that could be taken to ensure economy in using our national assets. What we clearly need to understand is that logistics has its roots in national economy, its technological base, industrial strength and the will of the government to accept the fact that defence helps a nation's all-round advancement. It is time we realised that by sharing knowledge and expertise we can establish a joint sector wherein the military needs and civil requirements can be met in an economical manner, thereby providing state of art technology to both the military and civil.

The level of science and technology and production efficiency in the civilian sector, including the private sector, indicates that it has the sophistication to effectively contribute to defence. The facilities and technological sophistication that are available in the non-military sector should be functionally integrated with the defence sector. Regrettably, the awareness in this regard, if it exists, has generally not been put to any constructive use. For creation of a 'national logistics grid' as a first step we will have to take our civil sector into greater confidence. The industry is willing to cooperate. Civilian industries need to be nursed till they become adept at handling intricate military requirements. One does not advocate that we should be reinventing the wheel. All that is required is a national policy on industry with identifiable defence usage where applicable and possible, and then task a core group to undertake the process of amalgamating "defence and development" in a time bound manner. The national goal should be to make the industrial capability of this country a realistic deterrent to any aggression, and a powerful addition to our war-fighting ability should deterrence fail. To achieve the goal of harnessing the national pool of resources we must establish a national policy and goal for mobilisation readiness. Throughout the political and military policy makers must pursue specific actions to support that goal. What really is required to be done is to take a holistic view of our civil needs and our defence needs in a pragmatic manner. Identify areas of commonality, identify those segments of our industry that can undertake defence-related projects, where required locate foreign collaborators and task specific groups to provide material support to the military. Such an action will ensure that the industrial capability could be built in such a way that it provides both the material for war and meets the needs of the country's population.

Broad Principles for a National Grid

In USA and UK where the decision-makers have had military experience (unlike Indian political leadership) every effort is made to integrate the logistics support to the services into the "national economy" so that the military needs are met quickly. They have largely focussed on integrating the defence needs with the economy and industrial growth. In the process certain broad principles have emerged. We could make use of the under mentioned guidelines. These are:

(a) **Central control of planning guidance**. All systems, types and levels of logistics planning are based on planning factors of one sort or another. The type used may vary greatly depending on the situation and the level of planning. In USA the 'Department of Defence Supplies' (DODS) takes all major decisions and provides to commanders at various levels, the guidelines for them to work out their logistical needs and how to go about ensuring appropriate material support within the accepted frame work. Since we do not have such an agency the task could be undertaken by the 'Planning Commission'. Chairman of the Chief of Staff Committee could ascertain requirements from all the three services, vet them, arrive at a common denominator and project the same to the commission. The commission in turn should relate these requirements to overall national needs and capabilities and allocate resources, industrial, financial and technological. Where the industrial capacity in the country is found wanting, either a foreign source be located, or where feasible domestic industry be encouraged to establish facilities.

(b) **Standardised Procedures and Equipment**: One key component of the international armaments strategy where civil facilities are put to maximum use is standardisation. In these countries the procedures and systems for working out material needs are standardised and well defined. The equipment also generally is common so that any procedural wrangling does not hinder the spare parts supply. Ammunitions are also generally of standard specifications, be it a Naval, Artillery or tank gun. The demanding drills and systems are also well defined. In the absence of standardisation logistics sustainment of the Military becomes much more difficult and complicated because as the quantities and types of equipment increase, so does the requirement of spare parts and qualified maintenance personnel. Spare parts become a problem because of the many varied makes and models and the resultant lack of interchangeability among their parts. Other than specific weapon and equipment systems we should encourage commonality to the maximum, particularly where vehicle fleet, transport aircrafts and such like things are concerned. Today too much is wanting in our system. We have more than twenty types of rifle 7.62 SLR in service. The confusion and consequent retardation in service to the troops such lack of standardisation can easily be imagined. But this is not the kind of arrangement we should live with any longer. In the East European systems the level of standardisation is such that an 'air compressor' after doing the prescribed hours in an air craft is installed in a bus or a tractor and thereafter, in a machinery item so as to derive maximum life out of the component.

(c) **Standardise Nomenclatures**: To make the task of the logistician even more difficult in our country; there are different procedures and systems within each service. Even within the same 'Service' various supporting services have different norms while working out their requirements. Strangely, certain items, even though they are the same in every respect, have different identifications because their usage differs. We have ' life jacket' of the Engineer's origin and ' life jacket' of Ordnance origin. Ostensibly, as the name suggests when the 'corps of Engineer's uses the item it has an identity which is different from other users. Both of them are expected to take provision action separately. During all his years in service the author could never understand how the holders of such like stores could really differentiate between the two. Another thing is that we tend to allot a separate index number to the same item whenever there is a slight change even in its packing and packaging, not an uncommon thing where an item could be coming in from different sources. This, besides being wasteful, leads to avoidable accounting problems at all levels. Therefore, there is a need to evolve standardised procedures and systems.

(d) **Combined Working of Liability**. The provision liability in respect of common items is worked out collectively. This, besides making the quantity on demand lucrative for the supplier, leads to competitive pricing as well. In our case all the three services do their own planning and provisioning even where the supplier is the same. There really is no point in Army, Navy, and Air Force to be separately going to the same vendors for the same items and paying much more than what they would have if the requirements had been processed collectively. This becomes all the more wasteful when the items are to be imported and the vendor could give a better deal if the volume was large enough. There is a need to set up a common body for processing indents pertaining to all the services. Where possible we should take the civil population's requirements also into account. More so, where a common usage product, procedure, system or processes are involved.

(e) **Avoid Duplication**. Existing facilities are to be used by all the customers, be they any branch of the military. Unless some aspect of security and secrecy is compromised even civil sector be allowed to make use of these facilities. In a limited way we are already following this principle where civil aviation is making use of "defence airports" and airforce stations. This kind of commonality is required to be extended in as many areas as possible. This helps in economical and practical utilisation of national resources. In a large number of cases, in order to establish priorities in meeting demands and to ensure a smooth supply, various agencies both in the defence and civil tend to duplicate sources of availability even when there is no scarcity or shortage of a particular product or service. Consequently, considerable amount of national resources is expended in certain activities that could have been put to a better use. It is difficult to understand why the Services should have their own vehicle repair workshops and overhauling facilities for commercial vehicles which are in common use even in remote areas of the country. Instead of making use of these facilities created by reputed suppliers of international standard and reputation we tend to create our own setups even when these have proved to be expensive and not up to the mark. Any number of studies has been carried out on the subject of 'base repair' and overhaul'. All these studies have established that more often than not professionalism is lacking in these base workshops. There are delays and slippage where customer satisfaction is concerned. Wasteful expenditure continues to be incurred for reasons other than operational merit. Logically, and if we want to economise on our defence expenditure, the Ambassadors and Leylands, TMBs etc., where possible; should be taken to the authorised service stations. There certainly will be a need to have repair workshops in remote and operational areas, but a reduction in these facilities in big towns can make army tradesmen available in a greater number where they are needed more. For that matter why should the Services be manning huge petroleum depots in places like Delhi or Mumbai. This leads to waste in two ways; firstly, every user unit has to go to the depots to collect their requirements periodically. Secondly, a lot of clerical effort is involved in accounting and facing a large number of "audit objections". A cheaper and economical way would be to give 'periodic service' and 'petrol' coupons to the users and they could go to any authorised service station to obtain their requirements. Adopting such a practice will ease a lot of pressure on the integral resources. Savings in these spheres alone can be economical to the state. In the process, involving the trade and the industry will make it easier for the Services to receive what they need without over-burdening the taxpayer. We must accept the fact that, blending civil needs with defence needs provides an element of participation to the citizens and financial elbowroom to the government. Therefore, duplication should only be selective and only where essential. As a broad principle one can say that it should be confined to only those areas where it is considered to be an operational necessity and is required as insurance or in order to create an environment of competition so as to enhance national efficiency.

(f) **Coordinate Information Flow**. Coordinated information flow (together with the appropriate communication network) to facilitate economic cross utilisation of inventories etc., and the collection and distribution of management information is necessary in order to optimise our resources. This is particularly so where procurements from foreign sources are involved. It is not unusual for the Army to be looking for a particular assembly pertaining to a weapon system desperately, even from abroad, when the same is held surplus with a sister Service and could be even under disposal as 'no longer required'. There have been cases where for want of information Navy has disposed 'anchor' of a decommissioned ship as scrap while ships of that class were in civil usage in India and even in some third world countries. (In Central Ammunition Depot Pulgaon a "loading machine" for L-60 gun was sent to the salvage area for its disposal as metal scrap because the weapon having gone out of Army service a long time ago this particular item had no use in so far as we were concerned. On the 'off chance' of finding its usage amongst sister services on raising a query it was found that the Navy was desperately looking for a loading machine for their guns. The depot pulled it out of the scrap yard and issued to Navy against 'book debit' and felt happy on having saved some money for the state. This happiness was, however, short lived because the "audit authorities" felt that Army should have charged Navy the current price and not the price at which it was procured more than twenty years ago.) The number of such like instances establish a fact that there is an urgent need for exchange of information on inventory related issues. We must involve other government agencies and even civil agencies when undertaking costly purchases or disposal of common user items. For major expenditures involving foreign exchange, setting up a joint purchase committee will be a big help. In the case of raw material like special grade steel etc., a national level purchase committee needs to be set up so that the purchases are regulated in a manner that takes into account requirements of all the potential users and places a bulk order on the supplier. The author on a visit to Vishakhapatnam where both Hindustan shipyard and the Naval dockyard are located learnt that they both had procured a ten tonne bale each of Korean steel separately when they actually needed four to five tonnes each respectively. Consequently, they both were now holding surplus steel procured with scarce foreign exchange and had no use for the item in the foreseeable future. Good economic sense should have meant a combined order to the foreign supplier. On a query as to why that did not happen it was told that since both the establishments were under different ministries they couldn't possibly have jointly processed the demand. It is such exasperating absence of logic that comes in the way of ensuring economy while dealing with the national resources. As a taxpayer one finds no reason for not exerting any foreign exchange saving measure, particularly when the two establishments are located only a few kilometers apart and are in the same business. There is a great need for a single coordinating agency that should have an efficient information system so as to make meaningful use of information with regard to deployment of resources, particularly, scarce foreign exchange.

An analysis of the discussion, so far, brings us to a conclusion, that as a poor third world nation we need to:

(a) Emphasise on integration of several facets of military logistics into a system that takes into account both the civil and the military requirements before deciding on a policy with regard to provision of goods and services.

What is meant here is that the military logistics system should be in harmony with its supporting economic system. There is no point in going in for grandiose plans of purely military significance if the country cannot afford them. Wherever such a gambit has been tried it has failed to generate any particularly higher level of confidence in so far as making a country strong and technologically powerful is involved. The use of commercially available equipment greatly simplifies the procurement of military equipment. This does not mean that all commercial equipment is suitable for military needs. Quite the contrary, much of it is unsuitable for the specialised demands that the combat operations make. For example, commercial light bulbs may not stand up under the shock of gunfire during actual operations. One will have to make a very judicious analysis of the military need and where defence can manage without holding on to reserves of commonly available items. That is to say, for the greatest effectiveness and efficiency the military should make the maximum practicable use of civilian equipment and techniques so that utmost economy is exercised in making use of the national resources. Defence needs a proper interface with the civil sector to achieve its operational capability at an affordable cost.

(b) Emphasise on Systematic Analysis Defence Logistics is not a requirement that exists in isolation or can be met without considerations that affect the nation as a whole. An analysis for the design of distribution systems will be called for so that man made shortages are not created, particularly when there is either a war or a national calamity. For every major national resource we should identify pure defence needs and civil needs. The defence should be given its share based on forecasts and actual requirements within that forecast provided budgetary provisions exist. If all these points are satisfied and there is not enough of that resource to meet civil requirements an effort should be made to locate a source. If necessary such a source be cultivated and nursed for a while so that shortages do not get created in due course of time. Thereafter, a complement be reserved for unforeseen which should also form bulk of "military reserves". If we think of creating large reserves then that means locked up capital without commensurate returns, plus the need to maintain these reserves in a suitable condition will mean an avoidable expenditure. For example, there is no need for the military to hold huge reserves of petroleum products under its own charge. The proposed national grid should look into this aspect and relieve services from the task of sitting on essential products, which has common usage angle. To keep things under control the position with regard to availability should be reviewed from time to time depending on criticality of a particular resource.

(c) Make Common Contingency Plans. As it is our military has always been called in to assist the civil administration whenever there has been a major disaster natural or manmade. It will pay rich dividends if our emergency plans with regard to disaster relief and procedures are also totally integrated with the country's defence needs. It may involve giving some type of legal and judicial status to the commanders at all levels but the common man will certainly be more confident of the system's ability to help him whenever there is a crisis. In America a special agency FEMA-Federal Emergency Management Agency was created in 1979 specifically to deal with all kinds of situations, including the aftermath of a nuclear attack. We could do the same as this arrangement will relieve military commanders from the worry of finding men to carry out above referred tasks and from carrying additional material for 'just in case' kind of requirements. All the equipments for emergencies enumerated above can be held in a central location to be airlifted to the area where a nuclear weapon has been used.

(d) Emphasise Systematic Management of the military logistics so as to integrate it into the overall goals of the country's security requirements, needs, aspirations of its citizens and industrial growth. Then only can we have a total national participation in providing a viable and economically sound defence-preparedness. For, logistics is not just the supplies of war materials that armies have on hand. It is also the ability of a national infrastructure and manufacturing base to support armed forces at war, the availability of national transportation means to ensure that those armed forces can be deployed, and the ability to resupply those forces when they have arrived in their areas of operation.

(e) Emphasise on Indigenisation Consider that any power that lags significantly in military technology, no matter how large is its military budget or how efficiently it allocates resources, is likely to be at the mercy of a more progressive enemy. With our present level of technological growth it should be possible to keep enhancing the degree of our self-reliance in a planned manner. In this we should in the first instance take those items and equipments which have potential for civil end usage. It should be possible for the government to encourage and even subsidise industrial units to introduce those items in the civil markets in addition to meeting military needs. A vibrant defence oriented technological base will generate need for more qualified personnel. The increased demand for scientists, engineers, electronic experts, and skilled technicians, stemming from both military operations and military research and development will lead to better employment potential for our young men and women. There are giant units like HAL whose capacities are grossly under-utilised. These units could undertake joint production and research with appropriate civil industry and improve the quality of their product. With the top order industrial houses buying from such organisations the product can find international acceptance as well. They could surely gain the confidence of international buyers and thus, take some of the "foreign exchange" load off our economy. However, to achieve this they will first have to gain the confidence of the Indian military and other major users in their products.

(f) Emphasise on Technological Integration. The facilities and technological sophistication that are available in the non-military sector should be functionally integrated with the defence sector. This is on the premise that whereas force deployment plans and other related subjects are the defence secrets, the planning of production and supplies is not treated so anywhere in the world. It has been established that functioning of defence units in isolation has largely been responsible for production delays and escalating costs leading to situations where we had to resort to imports when our production units already were working in the civil sector and had spare capacity to undertake defence requirement as well. The greater the economy achieved in military logistics, the greater will be the effectiveness, which the combat forces can develop within the basic limitations of the country- be it technological, financial or industrial. Diagrammatically, the sequence of actions can be represented in the following manner:

Diagrammatically, the sequence of actions for a 'National Logistics Grid' can be represented in the following manner

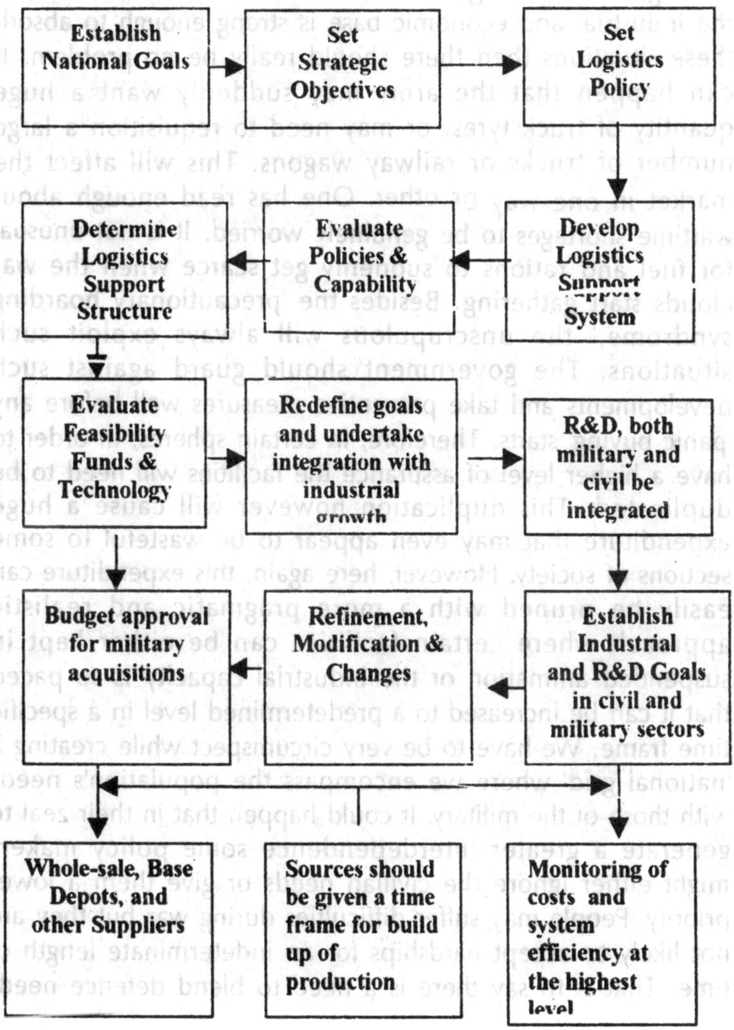

One aspect which can influence the decision with regards to integrating defence needs with the total production and transportation capability in order to provide appropriate and adequate logistical backing to the armed forces without seriously affecting the common man is the likelihood of shortages to the population as such. It can happen that defence needs could lead to some real or man-made shortages of certain goods and services for sometime. But if the industrial and economic base is strong enough to absorb these situations then there should really be no problem. It can happen that the army may suddenly want a huge quantity of truck tyres, or may need to requisition a large number of trucks or railway wagons. This will affect the market in one way or other. One has read enough about wartime shortages to be genuinely worried. It is not unusual for fuel and rations to suddenly get scarce when the war clouds start gathering. Besides the 'precautionary hoarding syndrome,' the unscrupulous will always exploit such situations. The government should guard against such developments and take preventive measures well before any 'panic buying' starts. Therefore, in certain spheres, in order to have a higher level of assurance the facilities will need to be duplicated. This duplication however will cause a huge expenditure that may even appear to be wasteful to some sections of society. However, here again, this expenditure can easily be pruned with a more pragmatic and realistic approach where certain facilities can be either kept in suspended animation or the industrial capacity is so paced that it can be increased to a predetermined level in a specific time frame. We have to be very circumspect while creating a 'national grid' where we encompass the population's needs with those of the military. It could happen that in their zeal to generate a greater interdependence some policy makers might either ignore the civilian needs or give them a lower priority. People may suffer difficulties during war but they are not likely to accept hardships for an indeterminate length of time. That is to say there is a need to blend defence needs with those of the populace in such a manner that the normal functioning and movement of industry & goods is not hampered in any way, unless there is an emergency. Even then it should be an exception that the civilian is made to change his manner of living due to a military exigency, and that too for the shortest duration of time.)

We have been rather slow in realising that in order to have an appropriate level of security without spending too much money, defence production and production in the civil sector would need to be augmented significantly for very valid reasons. Namely, to reduce our vulnerability to external pressures, to conserve much needed foreign exchange, to increase employment opportunities for our semi-skilled, skilled, scientific and engineering manpower and to gain competence in high-tech areas that would serve to upgrade other sectors of our industry. If these benefits are to be derived, a carefully thought out policy for developing defence industries and those civil industries that contribute to defence has to be formulated as an important element of national industrial, scientific and technological policy. Those technologies that improve the quality of life for the common man and also help bring in self-reliance in the field of defence need to be identified and nursed so that they remain competitive and contemporary. There will always be a need to evaluate the influence of any proposed step, whether economical or political on combat effectiveness of the troops once the national defence gets associated with a particular proposal for an industry or for an emerging technology. It could happen that a powerful industrial house corners the market and decides to hold both, military and the community to ransom. It could also happen that some preplanned export commitment clashes with an unforeseen defence requirement at a critical time. When such a situation arises the highest level of decision making has to involve itself in order to come up with an appropriate solution. To create a national logistics grid we can form a national level logistics council with the Defence Minister as the chairman.

Alternatively, Deputy Chairman of the "Planning Commission" can head this council. All the three Services should be represented on this council. The council should have representation from the finance ministry, Industrial Development Board, Department of Science &Technology, DRDO and representation from CII and FICCI. The council, concurrent with the 5 years plans, should do Perspective Planning for logistics. We should integrate those logistics disciplines that are common to all the three services. Where items are common and imported this should be done as early as possible.

Very broadly the important subject of "integrated logistics for the military and the civil population's needs" has been discussed. One thing, however, is obviously clear. That is something is definitely wanting in our present logistics support system. What is more difficult than getting a new idea into the military mind is getting the old one out. Now that we know that our present system of providing logistics for our military is outdated, and is out of tune with the technological developments it is time that we sat down and took those decisions that help our military to go into the next millenium with its eyes open and ears to the ground. Then only we can arrest our mounting "defence budget" and ensure greater functional efficiency in keeping our troops supplied with the wherewithal to carry out their designated tasks and role. This appears to be the only way the military can take its 'rightful place' in the realm of national development. This means that we have to take a look at our military requirements as a part of total national perspective where growth and development are concerned. The population need not be expected to suffer shortages and privations just because we are reluctant to accept that by involving various sectors of our industry and civil technology we will not necessarily be compromising on our security. The benefits of a total involvement and perspective are tremendous. Creation of a national logistics grid will be a step that can assure better military preparation and an improved quality of life.

To sum up we should quickly undertake the following:

(a) Identification of the capabilities and limitations of the national industrial base with regard to its ability to meet the entire range of technological demands of the Military.

(b) Active exploitation of indigenously available R&D and technology for in house development and production of 'state of the art' weapons and other military equipment.

(c) Formalise a development plan in various fields of those critical technologies that help both the military and industry.

(d) Prepare a blue print to build up progressively the technical manpower, particularly in those fields that have been identified as critical.

As the country rushes towards the 21st century, the Indian Army confronts difficult challenges and exciting opportunities. Advancement of technology and induction of latest equipments in the three Services is envisaged. The processes of rapid technological substitutions will not only lead to enhanced capabilities but may also gradually turn our military into a technocracy where combat to support personnel ratio may decline. How the Support Services transform themselves to meet these challenges and seize opportunities will determine the character of the 'Logistics Support System'. Will they measure up to the task is what concerns the operational commander at all times. The decisions and judgments we make today will form the guide rails as we journey into the future. We have a road map for that journey, but the elements are not necessarily in tune with a smooth journey. Frankly, they never are. At the end of the day it is your innovation, judgement of the situation and professional acumen that allows one to deliver the goods. We being products of our experience need to apply our minds to the subject of logistics if we wish to have successful operational plans in concert with national growth, technological advancement and industrial development.

Notes and References

1. *Affording Defence*, by Jacques S. Gansler.

18

Kargil Experience

Nursing assistants and doctors battled to save lives 600 meters below the peaks that the Indian army fought for. They would dash, scramble, fall and stumble up the slopes to evacuate the injured and dead soldiers. Many a time they were compelled to operate using a torch. Courageous drivers steered heavy trucks through winding mountain roads at night. The monstrous Bofors guns pulled by Scania trucks, Shaktiman, Ashok Leyland Stallions and hundreds of army vehicles loaded with food and ammunition would slowly make their way to various supply depots in the war zone. All this in the face of intense enemy shelling of the road. -(A press report.)

Twenty-eight years of near peace and only 'low intensity conflicts' and counter insurgency operations to contend with had made our military planners somewhat complacent about the support services. So much so there have been plans to further curtail the logistics tail by reducing manpower and resources employed therein. There were plans to effect reduction of as many as 50,000 men from -what is termed as 'non-field force' - so as to generate funds for modernisation. So much so, reduction in "Pioneer" companies, animal transport and certain other service elements has been under active consideration when Kargil happened. This sector has generally been guarded by a brigade and materiel resources have been commensurate with a 'brigade level' deployment. Continued tranquility and hardly any transgression by the

Pakistan army in this area since 1971 led to a situation where it was felt that the logistics elements needed no specific improvement *per se*. In any case the concentration of materiel elements in the region was more in tune with the requirements of neighbouring Siachen glacier. That, we were not ready to meet the challenge will be an understatement. The successful conclusion of the operation is a tribute to our combat soldier and his grit. This military success has come in the face of heavy odds. Not many experts believed that the campaign to evict well -entrenched Pakistanis from heights over 5,000 meters would be concluded successfully in a reasonable timeframe. That, this was possible for the combat soldiers to regain Indian Territory inspite of limitations of resources in the initial stages has been due to the logistics units and agencies rising to the occasion. As per the media reports as many as 12,000 rounds of artillery shells were fired in a day employing over 300 guns. The sector saw induction of almost three divisions in double quick time. Arrangements to provide logistics cover to such huge concentration of troops in active combat at heights ranging from 14000 feet to 18,000 feet could not have been possible without considerable effort on the part of our service units. The airforce played a very important role in ensuring a speedy build up by ferrying material from all over the country. The situation for the Ordnance became particularly demanding because the 'ammunition point' holding reserves for the brigade was blown up in the initial stages of the operation. With the Srinagar- Leh road under intense enemy shelling and helicopters (another means of providing logistics support) becoming dangerous to fly close to the supported troops due to Stringer missiles the task of providing replenishments of stores, rations, and fuel was made all the more difficult. To have succeeded in measuring up to the task under these conditions was no mean achievement. It will be appropriate to examine those factors that contributed towards provision of effective and reliable material support for the operation. Normally, the road to Leh from Srinagar is opened for 'winter stocking' in June every year. This year the stocking had not even started when the operation Vijay broke out and the Drass- Kargil-Leh road came under active enemy shelling making movement of vehicles virtually impossible unless India was willing to accept casualties of men and material. In fact, Pakistan had plans to cut this road axis so that the whole sector (including Siachen) could be eventually brought under its control. The aim plus was to raise the prospect of a nuclear conflagration and then use these developments to force a western meditated settlement. That it failed in its plans is a fact of history. Infantry, Artillery and 'combat support arms ensured this. In this they were actively assisted by all the support services. The Border Roads Organisation managed to pre-pone opening of the 'alternate route' via Manali- Rohtang Pass. This helped in sustaining the road movement of men and materiel. As many as 350 plus special trains carried troops, over a hundred trains carried materiel to the 'rail-head'. The railways deserve credit for making it possible. Thereafter, these stores were moved by road, by air, by animal transport, pioneers and porters to the forward localities. The men were airlifted by the Indian Airlines. This operation, truly, was a national effort in every regard.

This kind of a build up of men and materiel takes time. Therefore our existing system of holding "War Wastage Reserves" (WWR) and other reserves for an "interim period" i.e. reserves held to cater for the time taken for essential stores to fetch up paid huge dividends. The formation in Leh holds stocks of all types of stores including fuel, rations and ammunition for a period of eighteen months. The brigade at Kargil had limited stocks only for sustenance during the road-closure months and that too for maintaining a defensive posture where hostilities of this magnitude were not envisaged. With the enemy having brought road axis under its artillery fire it was not possible to supplement the dumps in Kargil-Dras area without taking heavy risks. These risks were taken and the materiel moved to the locations wherefrom the combat could be sustained. In this the reserves held in Leh and Srinagar came in handy in providing support to the troops in the initial stages. Uncommitted reserves from elsewhere could be brought in thereafter. Where artillery guns and other weapon systems were concerned the spares backing was provided by cannibalising guns held in reserve in depth. Ammunition, oils and lubricants were either 'para-dropped' or conveyed by road up to a specific point and 'man carried' thereafter. In this hiring of locals as porters played a major role because they are hardy and sturdy and required no or little acclimatisation. Equally effective were the 'pioneers' and the mules that carried heavy artillery shells to the gun locations.

The support system followed in this kind of terrain and operation is a combination of "Push and Transportation" model (discussed in earlier chapters). The advantages are: -

(a) The system aims at delivering the anticipated requirements of stores right at the doorstep of the user. It also emphasises a policy of delivery of stores forward to the farthest practical point in the combat zone.

(b) Commander can hold reserves centrally and can send them to locations where most needed. This reduces multiple handling/transshipment and therefore saves on time required.

(c) It relieves the forward commanders and staff from being tied down to send representatives and communications rearward for obtaining materiel and services.

(d) The lead-time (time between demand and supply) is compressed because the logisticians keep anticipated requirements ready to move forward. This is particularly important in the case of heavier ammunition that is kept on wheels for dispatch forward on receipt of requirement from a particular location.

(e) Accurate and timely information helps in prepositioning stocks at various echelons. All kinds

> and means of transport can be effectively exploited if the commander has the necessary inputs well in time. (In the instant case "multi model transportation system" was employed where airforce played a major role in taking essentials to Leh and Srinagar. The bulk stores were broken and packed in such a manner that minimum time was taken at every transshipment point).

Organisationally, all the procurement agencies were integrated, enhance financial powers were delegated to functionaries at various levels for speedy procurement of stores so that no time was wasted in making materiel available. Forward Surgical Centres from 'non-committed' formations were pulled out and deployed in the area of operation. Meals Ready to Eat (MRE) were made available to the troops and composite rations brought from other areas to supplement the ration requirements of fresh inductions. Snow clothing items and extreme cold clothing items were not enough for the forces deployed. This short fall was largely met by bringing in reserves from the base depots and from other theaters where there was no immediate requirement for the same.

Lessons Learnt and Suggestions for Improving Logistics Support

Retrospectively considering all aspects it is clear that holding of reserves in the region that remains cut off from rest of the country for more than six months has paid off. So has the expense incurred in creating an alternate route to Leh. The Army is making preparations to deploy division strength of troops (about 15,000 soldiers) in this uncharted and inhospitable mountainous terrain. In this sector several mountain ranges (15,000 to 18,000 feet) run north south along the loc. which means there are numerous valleys, nullahs and other gaps between the ridgelines. These gaps can serve as potential infiltration routes and will require monitoring and defending. It is neither economical nor practical to man every height in Kargil. Besides requiring a large concentration of troops we will also have to maintain them in extreme cold, and yet be unsure of any ingression of the enemy who can always sneak in. Therefore, it will be more pragmatic if high technology is employed in addition and the forces provided with the latest imaging and sensing equipment to cover gaps that will always exist. Local population can play a major role if they are given permanent employment in our regional depots and units. Squads of Patrol dogs will help in covering whatever voids are left due to terrain conditions. Following merit consideration urgently: -

(a) In an interview the COAS seems to have said that there is a need for better integration of resources at all levels, particularly at the National level. Also, requirement for a dedicated organisation at the national level to integrate defence needs into the country's development plan exists. (How it can be done has been discussed in the Chapter on "National Logistics Grid).

(b) Now that Kargil will have to be manned round the clock or at least the surveillance level augmented means that considerable increase in defence expenditure will be incurred. We can reduce the gap between the financial requirements and availability by integrating some of the common functions of the three Services. That is, creation of a 'Defence Logistics Corps' needs immediate consideration. There is no merit in having separate vehicle repair workshops, stores depots where common items are concerned and having fuel and ration depots. Integration of these services in peace locations can be started as a first step and we can prune these 'service elements' as we go along and gain experience in providing an 'integrated service'.

(c) Pioneers and Animal transport have proved their utility; so have certain other elements considered redundant earlier. Downsizing is definitely dangerous, particularly when it takes almost four years of training and grooming a 'mule' for an active operational role.

(d) There is an urgent requirement to improve means of communications in this region.

(e) As per the COAS (and certain other experts), India could have reduced its casualties if certain important items of equipment, particularly gun tracing Radar had been made available. This means that the bureaucratic slumber must give way to more energetic response to the military requirements in future.

(f) In our hurry to make Army 'lean and mean' a large number of service elements have either been eliminated or reduced. There is an urgent requirement for re-appraisal.

(g) It is neither desirable nor economical to keep importing recurring requirements of things like snow scooters etc; the R&D must be geared to come up with indigenous alternatives. We should integrate the civil sector also in this venture. Given a firm order they can come up with indigenous alternatives fast enough.

(h) Since all the assaults were platoon and section level action evacuation of casualty caused an additional strain on Infantry. In that, for every wounded soldier two to three men were required to carry him to the nearest medical aid post. Restoration of the "stretcher bearers" to the Infantry battalion should be given immediate consideration. As this consideration. As this operation has proved this was a misplaced economical measure.

(i) Logistics elements in the region need to be upgraded. In that, the Divisional Ordnance Unit and the Supply depot should be converted into "Regional Depots". Similarly, EME complement also requires certain additional tradesmen in various disciplines.

(j) All Terrain Vehicles (ATV's) play an important role in this kind of terrain. We should induct these vehicles quickly in adequate number.

(k) Snow and extreme cold clothing items were not available in sufficient quantity. Ordnance factory products are bulky and are of poor quality. Immediate needs are being imported. As a policy we should involve the private sector to make these items available to the troops in near future. In addition, we should adopt the local techniques for absorbing these climatic conditions. (Things like goat skin moccasins, sheep skin coats can be tried out in the case of those troops not actively deployed.)

(l) Most of the damage to the ammunition holdings and other logistics installations was caused due to shelling by enemy artillery. Gun Tracing Radar and other allied equipments like imaging sensors and night vision glasses are urgently required to prevent recurrence of this loss of essential stores which were fetched up on these heights with considerable effort and at a heavy cost.

(m) There is a definite requirement of building heli-pads and airfields in the immediate vicinity of our forward depots and store holding units.

(n) Most of the civilian truck drivers were reluctant to take their vehicles to the battle areas. Persistent enemy shelling and consequent damage to the trucks and injury to the drivers/cleaners dissuaded truck owners from letting Army hire their vehicles for conveying stores from rail or road head to the forward areas. Army can consider providing insurance cover to all those involved in assisting military during active operations. Alternatively, under the aegis of Director General Resettlement a Road Carrier service consisting of ex-servicemen can be started in the entire region. This agency should carry stores from the rail/airhead in normal times as well so that this becomes a whole time employment for the ex-servicemen.

(o) ISRO will be launching a surveillance capable satellite next year. This will provide Indian Military with a powerful means of monitoring activities of not so friendly neighbours. This satellite (Cartosat-1) will represent India's first photoreconnaissance vehicle has been developed for defence intelligence analysis. It will be essential that any information received/gathered through this medium be quickly passed on to the commanders on ground. The usual bureaucratic channels where considerable time and effort gets wasted- as it happened recently when precious time was wasted in finding the actual intentions of the enemy and consequent delays in building up resources- can be avoided.

(p) There will be refugees and other locals who will require rations and medical aid if injured due to enemy shelling. Since resources in the immediate battle zone cannot be expended on them without having an impact on the operation it will be better if they are evacuated to the rear in returning vehicles. This will also help in reducing pressure on the 'road space', which gets affected when the local reaction is unorganised and unmonitored in such a situation.

Political and bureaucratic propensity for knee-jerk reaction to an operational situation is too well known and extensively catalogued by various experts to merit repetition. Need of the day is to quickly overcome the mental blocks and create an environment where military can function effectively in response to fresh challenges that Kargil has thrown up. Government has sent abroad officers who were actually involved in recent fighting to look for equipment the military needs for high altitude warfare. This is a welcome development provided the same is quickly bought and inducted into service.

Bibliography

1. Technology and War from 2000 BC To the Present by Martin Van Creveld
2. The Complete Strategist by J D Williams Published by McGraw Hill
3. Logistics in the National Defence by Henry E Eccles
4. Defence Management by Stephen Enke
5. Logistic Management by James R
6. Military Logistic Systems in NATO by Geoffery Ashcroft
7. Powering War- Brassey Publication
8. Economics of Defence by Dr. Jai Narain
9. The Organisation of Logistic Support Systems Edited by R Bureau
10. Logistic Support- Vietnam Studies by Lt. Gen. Joseph M Heiser, JR
11. INDIA'S ADHOC ARSENAL -Direction or Drift in Defence Policy? By Chris Smith
12. The Sinews of War: Army Logistics by James A Huston

13. The Economics of Defence by Todd Sandler
14. Moving Mountains Lessons in Leadership Logistics from the Gulf War By Lt. General William G Pagonis
15. Technology in War -by Kenneth Macksey and A Basch, Rinehart & CO. New York
16. War and Defence Economics by J Blackman
17. Logistical Support of the Armies by Roland G Ruppenthal
18. War, Economy and the Military Mind edited by Geoffrey Best
19. Defence & Economic Growth in Developing Countries by E Benoit,
20. Supplying War by Martin van Creveld
21. The Defence of the Realm in the 80s by Dan Smith
22. The Impact of New Military Technology Edited by Jonathan Alford
23. Affording Defence by Jacques S. Gansler
24. Resource Management in Peace and War by Louis C. Seeling
25. Defence versus Development by D D Khanna and P N Mehrotra
26. Defence Without Drift by P.V.R Rao
27. The Effects of Nuclear War - published by Office of Technology Assessment -US

28. The Effects of Nuclear Weapons edited by S. Gladstone and PJ Dolan (US Department of Defence and Energy publication).

29. Effects of Nuclear War, Journal of the Federation of American Scientists, 34(No2 1981)

30. Nuclear Weapons: Principles, Effects and Survivability by Charles S Grace Royal Military College of Science UK

31. Tactical Nuclear Weapons: An Examination of the Issues by William R. Van Cleave and S.T. Cohen

32. Future Wars by Trevor Dupuy

33. Conventional Warfare in the Nuclear Age by Otto Heilbrunn

34. Nuclear War-Survival Skills by Cresson H. Kearn

Index